TEXAS A&M UNIVERSITY PRESS
COLLEGE STATION

Texas Rattlesnake Roundups

Clark E. Adams & John K. Thomas

LIBRARY OF CONGRESS CATALOGING-IN-PUBLICATION DATA

Adams, Clark E. (Clark Edward), 1942–
    Texas rattlesnake roundups / Clark E. Adams and John K. Thomas. — 1st ed.
        p. cm.
    Includes bibliographical references and index.
    ISBN-13: 978-1-60344-035-6 (flexbound : alk. paper)
    ISBN-10: 1-60344-035-6 (flexbound)
    1. Rattlesnake Round-up—History.   2. Rattlesnake
hunting—Texas—History.   3. Rattlesnakes—Texas.
I. Thomas, John K.   II. Title.

SK341.S5A33   2008
639'.14—dc22   2007048038

Contents

Acknowledgments   vii

Introduction   ix

Chapter 1. Anatomy and Natural History of the
    Western Diamond-Backed Rattlesnake   1

Chapter 2. History of Texas Rattlesnake Roundups   25

Chapter 3. Organization of Texas Rattlesnake Roundups   33

Chapter 4. Impact of Hunting on Rattlesnake Populations   67

Chapter 5. Conclusions   79

Appendix: Spectator and Hunter Interviews   91

Glossary   101

References   105

Index   109

Jackie Bibby demonstrating one of the rarer stunts conducted by snake handlers (used with permission from Guinness World Records 2006).

Acknowledgments

The information about rattlesnake roundups in this book involved the contributions of many people. The original research team included McAlister Dowd Maxwell, Janice Schnake Green, Stephen Jester, and Kelly Strnadel Scroggins, who all helped conduct personal interviews at selected roundup communities and processed much of the data. Ernesto Enkerlin conducted all interviews requiring Spanish translations at the Freer roundup. During the 2006 follow-up effort, Jessica Alderson assisted in conducting interviews and photographing the Taylor and Freer roundups. Regina Rohde developed several maps on roundup locations in the state, and Linda Causey provided drawings. The Sweetwater Jaycee organization and Kathy McGinty, a Texas Parks and Wildlife Department biologist, provided historical data on the rattlesnake harvests at Sweetwater. Robert Ackerman, Steve Raines, J. E. Morales, Jackie Bibby, Ken Darnell, and their associates provided important information on the status of rattlesnake roundups in 2006 from the perspective of snake handlers and dealers. We are grateful for the opportunity to have visited with them and for having shared with them the experience of observing what people do to western diamond-backed rattlesnakes at these roundups.

The information we provide is intended for a general audience—readers for whom this book is their first exposure to the Texas rattlesnake roundup phenomenon; people who may have attended rattlesnake roundups but want to become more informed about them; and those who may support or oppose rattlesnake roundups. To more effectively make the content of this book relevant to these audiences, we received editorial support from Judy Adams, Sara Ash, and Kieran Lindsey, each of whom represent one or more of the types of readers for whom this book was written. We are grateful for the time and professional expertise they donated to the production of this book.

Finally, this book would not have seen the light of day had it not been for the patience and perseverance of Shannon Davies, who encouraged and counseled us through a demanding external review process. We thank her and other members of the Texas A&M University Press staff who helped us to finally put a cover on a three-year effort.

Introduction

The primary focus of this book is the western diamond-backed rattlesnake, a member of the genus *Crotalus* and the species *atrox*. This taxonomic name may not mean much to most of us, but it gives clues about the unique perception people have of this animal. The Greek term *Crotalus* means "rattle" and the Latin term *atrox* means "cruel" or "horrible." The terms together refer to the "horrible rattler," or what most people know as the rattlesnake.

Texas has seventy-two species of snakes, eleven of which are dangerous to people. Two species account for the most bites inflicted on people, pets, and livestock—the copperhead (*Agkistrodon* sp.) and the western diamond-backed rattlesnake. According to the Texas Department of State Health Services, the western diamond-backed rattlesnake is the only venomous snake to have killed an average of one person per year in Texas from 1978 through 1995. It is also the only snake that people deliberately hunt so as to celebrate its extermination in an event known as a "roundup." Rattlesnake roundups have drawn much attention from wildlife professionals, social scientists, and special interests groups in the past two decades because of this unique interaction between people and the species.

Neither of us in our home states of Iowa and Mississippi encountered rattlesnakes while on camping, hunting, and fishing trips in our youth. Our experience with the western diamond-backed rattlesnake began professionally in 1990, when we first heard about rattlesnake roundups. A rattlesnake roundup is an annually conducted community fund-raising event in which live rattlesnakes are collected, commercially traded, and displayed in a carnival-like setting totally unlike traditional hunting of game. Our interest in rattlesnake roundups grew out of a biological and sociological curiosity about how people and their communities' activities affect this particular wildlife species.

Some Texas Parks and Wildlife Department (TPWD) biologists are aware of roundups and are concerned about the potential impacts these events are having on the western diamond-backed rattlesnake population. Rattlesnake roundups represent a hotbed of controversy between organizers and opponents, yet the TPWD leadership knows only that they exist and generally knows little else about the phenomenon. Having already conducted several studies of Texas hunters for the TPWD, we are familiar with many of the wildlife management issues the agency needs to address. But this issue is unconventional because rattlesnakes are a nongame species that is being hunted, and several Texas communities have institutionalized such hunts in the form of roundup events—in other words, roundups have developed and expanded over several decades as annual events until they have become, in some cases, an inherent part of a community's identity. Consequently, our discussion of rattlesnake roundups includes the impacts on rattlesnake populations as well as on the communities that promote the hunts.

From their beginning, rattlesnake roundups have been a contentious issue to professional herpetologists (individuals who study amphibians and reptiles) and animal rights activists. Our research is focused in part on the ongoing debate between opponents and proponents of Texas rattlesnake roundups. This debate is presented by addressing opposing points of view and identifying what issues need to be examined in order to verify or refute the arguments.

In 1991 we selected for study five of the seventeen communities in Texas with active roundups. The communities chosen represented geographically different areas of the state. Two of these communities had the largest numbers of spectators at their roundups—Sweetwater (North Texas) and Freer (South Texas). The Central Texas communities of Big Spring, Brownwood, and Taylor were also included. Except for Taylor, these communities are located in non-metropolitan counties. All the communities had less than ten thousand residents, except Sweetwater, and all had conducted an annual roundup for nineteen consecutive years or longer. Personal interviews provided harvest data from hunters and opinion and behavioral information from organizers, rattlesnake dealers, members of snake handlers' clubs, and public spectators. Many photographs, by us and others, documented the unique features, personalities, and events at each roundup studied.

Although much of the information presented in this book is based on the data acquired in the original investigation in 1991, a follow-up investigation was conducted in 2006 in which three of the seven surviving roundups were revisited—Freer, Sweetwater, and Taylor. After a fifteen-year hiatus, harvest and organizational data were updated, personal interviews with dealers and vendors were conducted, and more photographs were taken.

Phenomenal changes in the conduct and organization of rattlesnake round-ups had occurred over the fifteen-year period between studies. The technological changes alone provide some historical perspective. For example, in 1991 we took photographs with a 35 mm single reflex lens camera; by 2006 we had gone digital. In 1991 we relied on the print media and "snail mail" to access information about rattlesnake roundups. In 2006 we used email, cellular phones, and the Internet, which provided nearly immediate access to an information explosion on the topic. Nevertheless, rattlesnake roundups have continued, resulting in the permanent removal of tens of thousands of Texas western diamond-backed rattlesnakes from their natural habitats.

We drew information about western diamond-backed rattlesnakes and Texas roundups from many sources, which are listed alphabetically at the end of the book. Our account of rattlesnake roundups begins with an overview of the anatomy and natural history of western diamond-backed rattlesnakes. Next we present a history of the roundups conducted in Texas and a description of roundup organization and the characteristics of participants. We then discuss possible impacts of roundup hunting and unregulated commercial exploitation on rattlesnake populations.

We conclude with some general thoughts about the future of rattlesnake roundups in Texas, the unfulfilled need for the TPWD to determine the ecological impacts caused by roundups and other collection practices on rattlesnake populations, and recommendations about how a study of these impacts might be conducted. Boxed vignettes add stories about encounters between people and rattlesnakes.

Rattlesnakes have a special place in American folklore. Some of us may recall the "Saga of Pecos Bill" from our youth. Pecos Bill is said to have been born in Texas in the 1830s. According to the lore, he used a bowie knife as a teething ring and played with bears and other wild animals when he was an infant. After falling out of his parents' wagon near the Pecos River, he became lost and was subsequently raised by coyotes. As the legend goes, Pecos Bill rode a mountain lion and used a rattlesnake as a whip when he grew up to become a cowboy. No one knows whether this character existed outside the imagination of Edward O'Reilly, who wrote of Pecos Bill's adventures in 1923, but rattlesnakes continue to be an important part of our image of a wild and untamed Southwest.

No other group of animals evokes people's fear and misunderstanding more than rattlesnakes. If threatened, the western diamond-backed rattlesnake is considered the most aggressive rattlesnake in the United States, and it is one of the most dangerous due to its abundance and potent venom. It was the most feared snake in the United States, according to a 1999 Harris poll.

Our fear of snakes in general is even more prevalent in contemporary urban society, perhaps because of our detachment from the natural world around us. Richard Louv, in *Last Child in the Woods*, referred to this detachment as a "nature deficit disorder." Psychologists believe that we learn our fear of snakes and that we pass this fear on to our children. Human fear of snakes is grounded in mistaken beliefs that all snakes are venomous, evil, and aggressive. Human aversion to snakes stems from the perceptions that they are sneaky, cold to the touch, do not blink but always stare at you, stick their tongues out at you, and have a peculiar looking body. Snakes may evoke images of evil because people believe snakes have been cursed by God. After all, was it not the snake, the Devil's emissary, that tempted Eve to eat the forbidden fruit in the Christian scriptures?

Our fear of snakes is reinforced by what we read, what we see in old Western and adventure movies, and by what we experience, so it is little wonder that many people believe the only good snake is a dead snake. Once people learn some of the interesting facts about snakes and discover that most are harmless and beneficial, their aversion and the mystery surrounding snakes may diminish, and the impulse to kill snakes may become more restrained. But in many cases the fear never really goes away. Some people just seem to have an innate fear of snakes (ophidiophobia), the way others have a fear of heights (acrophobia) or closed spaces (claustrophobia).

Texas Rattlesnake Roundups

Chapter 1

Anatomy and Natural History of the Western Diamond-Backed Rattlesnake

Humans have to be impressed with the genetic design, the danger, and the unique behaviors of a western diamond-backed rattlesnake. It is difficult to imagine all the evolutionary forces that were necessary to produce a reptile with the remarkable adaptations that the rattlesnake possesses. Nevertheless, many people do not fully appreciate nature's designs, particularly when the lethal part of the design can result in disastrous consequences for human health and economic well-being. Ironically, it may be the lethal part of the design that shapes people's' curiosity and interest in rattlesnake roundups. The roundup may also symbolize the perception of human domination over one of nature's most dangerous wild creatures.

The following discussion of rattlesnake anatomy and natural history focuses only on those aspects that are connected, directly or indirectly, to roundups: the skin and rattle; the head; the strike, bite, and venom; hibernation; reproduction; diet; and danger to humans and livestock. We conclude this section by addressing several common myths about rattlesnake anatomy and natural history.

Skin and Rattle

The entire body of the rattlesnake is covered by a single sheet of skin that is folded and creased to form thickened and hardened scales. The arrangement of scales forms a flexible body covering that is both strong and durable. The diversity of colors and patterning on rattlesnake skin is considerable. Color patterns vary by species of rattlesnakes and include hues of gray, brown, pink and red. However, in Texas most western diamond-backed rattlesnakes are grayish brown or buff with rhombic or diamond-shaped blotches of darker brown along the back surface, fading as they extend down onto the sides (see fig. 1). The belly of the snake is cream or buff, and black and white rings alternate on the tail, up to the rattle.

Figure 1. The skin of the western diamond-backed rattlesnake is a beautiful array of colors and patterns as demonstrated in a close-up picture in the holding pit (by Clark E. Adams).

The rattlesnake's skin provides the protective coloration or camouflage that the snake needs to hide from enemies and to be less visible to its prey. Skin shedding (known as ecdysis) is required for growth and to replace worn skin. Rattlesnakes may shed as many as six times a year, with young snakes shedding more frequently than older ones. The covering of the eyes and the lining of the pits are replaced each time. In fact, hunters use the presence of shed skins, called "sluffs," to indicate the presence of rattlesnakes in an area.

The skin also registers air temperature, giving the rattlesnake the information it needs to seek shelter or a warm rock. Rattlesnakes are cold blooded (called ectothermic), meaning that their body temperature is influenced by ground and air temperature. The optimum temperature range during which most snake activity occurs is between 44 and 90 degrees Fahrenheit. This is why rattlesnakes are seen more often in the summer than in winter months.

Unlike mammals, reptiles grow throughout their lifetime based on food availability. As a result, the number of warm months that a rattlesnake has to hunt and consume prey makes a difference in size in different parts of the state. Mature western diamond-backed rattlesnakes in Texas vary in length from three to five feet in the northern part of the state and from five to eight feet in South Texas. It should be noted that an eight-foot western diamond-backed rattlesnake would be a rare find, and any claims of capturing one of this size would draw a measure of skepticism among experienced field herpetologists.

A photograph taken at the 1991 Freer rattlesnake roundup shows an extremely large snake (fig. 2). The mounted snake was the property of the snake dealer who was contracted by the roundup organizers in Freer in 1991 to provide live western diamond-backed rattlesnakes for exhibition. When captured, it was claimed, the mounted rattlesnake weighed seventeen pounds and was over eight feet long. After this photograph was taken, one of the roundup vendors, a local resident and school teacher, unrolled a tanned western diamond-backed rattlesnake skin that was three inches longer than that shown in the photograph. The skin measured twelve inches at its widest point. Both the mounted snake and the tanned skin had more than twenty segments or buttons in their rattles. One of the people reviewing our information questioned the ability of a western diamond-backed rattlesnake to achieve this size in the wild and pointed out that tanned skins can be stretched by up to 30 percent. This

makes it possible for both skins and mounts to exaggerate a snake's size. No one really knows the growth potential of the western diamond-back when allowed to complete its life cycle in its natural habitat without human activites interrupting the process.

The rattle is part of the skin and is composed of hollow, interlocking segments of keratin, the same substance found in horns, feathers, and people's

Figure 2. A taxidermy mount of one of the largest western diamond-backed rattlesnakes ever captured (by John K. Thomas).

fingernails. Each time the snake sheds its skin a new segment is added to the rattle. Shedding cycles depend on climate and prey availability. Warm weather and more abundant prey mean a greater food supply, which results in faster growth and more frequent skin shedding. Counting the number of segments in a rattle is not, therefore, an accurate way to determine the age of a rattlesnake. A rattlesnake could have added or lost several segments in the course of a year.

Herpetologists and others who have studied the structure of the rattle believe it evolved as a warning mechanism. It continues to serve this function when hiding and escape are no longer survival options for the rattlesnake. Thus it is unlikely that a rattlesnake would use its rattle when hunting for prey—it would be like attaching a bell around the neck of a cat. For unwary humans and other creatures too large to swallow, the message coming from the sound of the rattle is "continue at your own peril." We recall becoming so accustomed to the buzz of the rattles after visiting several roundups that we no longer had an alarm reaction when we attended the roundups held later in the spring months. Snake hunters and handlers say that the most dangerous rattlesnake of all is one that has lost its rattles (a rare occurrence) and cannot signal an alarm or one finding that its rattle is ignored by humans.

Head

The rattlesnake's head is equipped with everything needed to protect it from enemies and to identify, stalk, capture, kill, swallow, and digest its prey (see fig. 3). Although it lacks external ear openings and is essentially deaf, a rattlesnake uses other senses to detect threats and prey. Its eyes are lidless but are protected by a hard transparent covering that is shed with the skin. The western diamond-backed rattlesnake's pupils are elliptical (vertical slits), and sight is limited in distance and direction. However, the snake is also equipped to respond to thermal imagery by using heat-sensing pits situated between the eyes and the nostrils and oriented directly forward. These small openings detect thermal (heat) radiation in the environment. Thirty-five milliseconds elapse between reception of stimuli to the pits and a physical response, such as striking (one millisecond = one millionth of a second). Temperature signals are relayed to the same region of the brain that processes visual information, and a thermal image is superimposed on the visual one.

A rattlesnake has a keen sense of smell. The external nostrils can detect conspicuous odors of enemies and decomposing prey. The association between the tongue and a special organ in the brain, called the Jacobson's organ, serves as a chemical receptor system that is used to detect airborne particles. This mouth-nose system

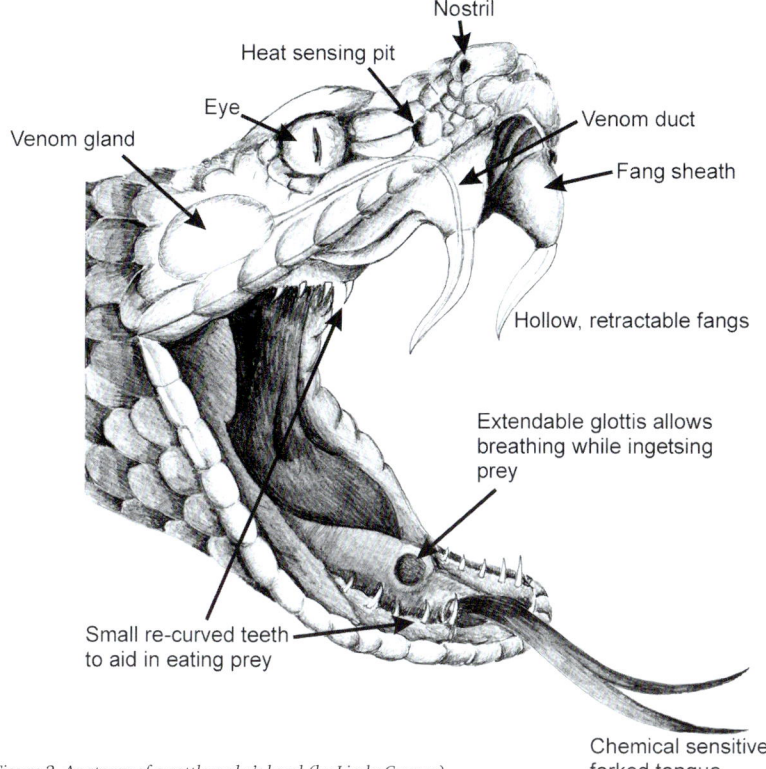

Nostril

Heat sensing pit

Eye

Venom gland

Venom duct

Fang sheath

Hollow, retractable fangs

Extendable glottis allows breathing while ingetsing prey

Small re-curved teeth to aid in eating prey

Chemical sensitive forked tongue

Figure 3. Anatomy of a rattlesnake's head (by Linda Causey).

operates when the tongue picks up particles and gases from the air and is retracted into the mouth, where the tips are brought into contact with pits in the Jacobson's organ. This system of smell is vital in locating prey and potential mates.

Strike, Bite, and Venom

Venomous snakes differ in the location and operation of their fangs and the composition of their venoms, and taxonomists use these characteristics to separate the snakes into families. There are four families of venom-ous species. Snakes in three of these families have rigid fangs used to bite, inject venom (envenomate), and hold their prey until it dies. These families include the Colubridae, such as the African boomslang, which have rigid fangs in the rear of the mouth. The rigid fangs of snakes in the Elapidae family (e.g., coral, cobra, and mamba) are located in the front of the mouth. Members of the Hydrophidae, the sea snake family, have rigid front fangs and are completely aquatic.

The remaining family consists of the Viperidae (pit vipers), including adders, asps, rattlesnakes, and copperheads. All have hinged fangs that spring forward during a strike. A rattlesnake's fangs are hollow, resembling hypodermic needles in structure. They fold against the roof of the mouth when it is closed. These long, curved structures are rotated out and are pointed almost directly forward when the snake strikes. These snakes release their prey after envenomation, track it as the venom gradually incapacitates or kills it, and then consume the prey by swallowing it whole.

A fang is useful for six to ten weeks before it is replaced by another located behind it in the jaw. Fangs that are broken or lost during striking and feeding activities can also be replaced. A rattlesnake possesses up to seven additional sets of fangs in various stages of development behind the functional fangs. The fangs are connected to venom glands located at the back of the upper jaw. In the upper and lower jaws and behind the fangs are small solid teeth slanting backward, used to hold prey during the swallowing process.

After potential prey is located, the snake lunges forward with its mouth wide open (about 180 degrees) with fangs pointed forward. Striking is primarily a stabbing motion followed by an arching of the neck to insert the curved fangs. Eastern diamond-backed rattlesnake *(Crotalus adamanteus)* strikes have been documented at speeds of 175 miles per hour!

Because of this striking motion and the location of the eyes, a rattlesnake does not actually see its prey or victim at the moment of envenomation. It is capable of striking at objects directly overhead and can strike at a distance of one-half of its total body length. Several factors affect the distance of the strike, such as size and species of snake, anchoring substrate, position of the snake before the strike, and the degree of a snake's excitement.

Size and type of prey can influence whether a strike is followed by a quick release or a holding bite. For example, northern pacific rattlesnakes *(Crotalus viridis oreganos)* release their prey after an initial strike in the head and thorax regions. Herpetologists believe this behavior prevents the prey from administering retaliatory bites and that larger snakes are more likely to hold prey after striking. In contrast, a western diamond-backed rattlesnake releases its prey after injection of venom and will track and locate the prey by using tongue flicks and its pit organs.

Venomous snakes differ in the proportions of hemotoxic and neurotoxic enzymes present. Rattlesnake venom is toxic saliva that includes a mixture of enzymes unique to pit vipers. Hemotoxic enzymes destroy blood vessel integrity and disrupt clotting mechanisms in the blood. To a lesser degree, other enzymes are neurotoxic and paralyze the nerves in respiratory and heart muscles of prey.

Venom has several functions. In addition to immobilizing and killing prey, it aids digestion by loosening hair and feathers, breaks down the body tissues of the prey, deters predators, and kills bacteria released from the gut of the prey. Venom is carried through the circulatory system of the prey and starts digestion. Within the rattlesnake's digestive tract, the enzymes in the venom help prevent putrification of large prey, which can take several days to digest.

One commonly held misconception is that a young rattlesnake's bite is more dangerous than that of an adult. Young rattlers, with their short, delicate fangs and limited supply of venom, are not in fact more dangerous but are as dangerous as their adult counterparts. What they lack in volume of venom in a bite is compensated for with a more concentrated form of venom.

Herpetologists and others who handle rattlesnakes have long debated whether a rattlesnake can actually control how much venom to inject into its prey. Some herpetologists contend that rattlesnakes can control at will the amount of venom discharged from either fang, from both, or from neither. Other herpetologists disagree with this control theory of envenomation. They suggest that the anatomy and muscular action associated with striking and biting dictate automatic envenomation, with the amount of injected venom being dependent upon volume of reserves in the venom glands. In a review of venom metering by venomous snakes, Young and associates concluded in 2002 "that there are multitudes of anatomical, physiological, and behavioral characteristics (whether viewed as internal or external factors) that could influence venom flow." For example, disease, genetic abnormalities in the fang/venom sac apparatus, physical damage to the fangs, or angle of the strike could affect the success of envenomation during a bite. Young and associates further concluded that it would seem preferable to interpret differential venom flow as resulting from physical interaction alone rather than as involving decision making.

These factors may explain the "dry bite" phenomenon reported by many rattlesnake handlers and others. A dry bite results when no venom is transferred to the intended victim. Another explanation of dry bite is that it may be the result of an odd alignment of the snake's jaw, preventing envenomation as it struck at an overlarge target. Such would be the case with younger, smaller rattlesnakes. Otherwise, a rattlesnake can sometimes simply miss striking its intended target, resulting in a glancing bite that does not completely puncture the skin (Jim Dixon, pers. comm.).

Some people are so terrified by snakes that the actual event of being bitten, even by a nonvenomous species, can be fatal. A person can become so distraught during such an encounter that fear alone brings on a heart attack. For this and other reasons, experts advise snake bite victims to remain calm but to take immediate and deliberate action to seek aid. We discuss aid in more detail later.

Habitat and Prey

Terrain does not pose an apparent problem for rattlesnakes seeking their prey. Rattlesnakes have been observed climbing rocky outcroppings, bushes, and trees in addition to crawling on dry land. Snakes will climb three to four feet high after prey and have been seen in creosote bushes, prickly pear, and mesquite trees. Although rattlesnakes do not swim regularly, they are capable of it and are known to swim to and from Texas' coastal islands. It is unknown why they would make these aquatic expeditions.

As already indicated, when rattlesnakes become hungry (feeding periods are several weeks to months apart), few obstacles get in the way of their pursuit of primary prey species, which include small mammals such as rodents and rabbits, birds, other reptiles, and amphibians. Baby rattlers eat some types of insects, such as crickets. Seven-foot rattlesnakes in South Texas are large enough to consume jackrabbits.

By harvesting their prey, rattlesnakes may provide an important ecological service in local food webs—they help to keep the size of rodent populations at tolerable levels for humans and ecosystems. Rodents can carry diseases that are communicable to people, such as plague and Hanta-

AN EARLY RECORD OF A DRY BITE

What seems to be a record of a dry bite episode by a venomous snake appears in the Bible:

But when Paul had gathered a bundle of sticks and laid them on the fire, a viper came out because of the heat, and fastened on his hand. So when the natives saw the creature hanging from his hand, they said to one another, "No doubt this man is a murderer, whom though he escaped the sea, yet justice does not allow him to live." But he shook off the creature into the fire and suffered no harm. However, they were expecting that he would swell up or suddenly fall dead. But after they had looked for a long time and saw no harm come to him, they changed their minds and said that he was a god. (Acts 28:3–6, New King James Version)

Some believe that the leopard snake (Elaphe situal leopardi), a native nonvenomous species, was the "viper" that bit the Apostle Paul on his arrival in Malta.

Source: http://www.geocities.com/RainForest/3096/snakes.html.

virus. As a practical matter, consider the fact that a single adult rattlesnake can hypothetically consume up to ten adult rats per year. An adult rat can easily consume 25 pounds of grain per year and can produce as many as ten litters in a year, with an average litter size of eight, which equals eighty more rats! Thus, by killing one female rat, a single rattlesnake could save a farmer an estimated 2,025 pounds of grain in a year.

However, the removal of that one rat may mean that other rats can reproduce more because of reduced competition for available food sources. A common lament among those who oppose rattlesnake roundups is an anticipated increase in the population levels of rats and mice when rattlesnakes do not eat them. Their concern is based on a fundamental population control mechanism in natural ecosystems called the predator-prey balance. This control mechanism can be examined by reviewing the basic principles of population dynamics (see How Animal Populations Increase, Decrease, or Remain Stable).

One of the major concerns voiced by those who oppose hunting rattlesnakes for roundups and thus removing them from their natural habitat is the assumed negative impact this removal will have on the predator-prey balance. This concern needs to be addressed by first examining population dynamics illustrated by the predator-prey balance.

Figure 4 is a hypothetical illustration of the predator-prey balance. To keep the example simple, assume that habitat conditions promoting prey population growth (biotic potential) are optimal and that other species are unimportant in the dynamic interrelationships between predator and prey. Note in figure 4 how the size of the prey population decreases as the size of the predator population increases. The more food there is to eat, the more likely predators will have offspring, and in greater numbers. The offspring that would survive to become adults would in turn reproduce, and their offspring would do likewise. Conversely, observe how the predator population numbers decline when the size of the prey population is low.

Given few predators, the number of prey will increase at or near exponential rates (e.g., 2, 4, 8, 16, 32, 64, etc.), eventually exceeding the number of animals that the habitat can support, known as the "carrying capacity." This explosive growth rate, illustrated in figure 5 using rats as an example, will soon cause them to consume all available food sources. This irruptive growth will thus, in turn, lead to serious habitat damage, causing a massive die-off of the prey population with consequent adverse affects on other plant and animal species occupying the same habitat. The cycle of the prey population exceeding the carrying capacity of its habitat followed by a population dieback will

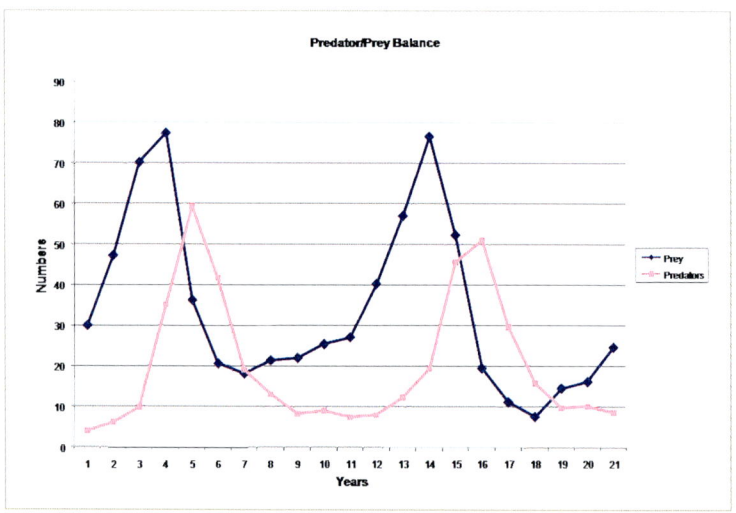

Figure 4. One example of a predator-prey balance that shows how the numbers of predator and prey species oscillate over a period of twenty years (by Clark E. Adams).

HOW ANIMAL POPULATIONS INCREASE, DECREASE, OR REMAIN STABLE

Wildlife managers study species in order to understand how many in the population represent a "surplus" relative to what the ecosystem can support. For example, they try to determine: (1) the population density, which is the number of animals in a particular area, such as the number per acre; (2) how quickly and how many offspring are produced by the population over a year—referred to as the reproductive or growth rate; and (3) whether the difference between the number normally present and number produced equals a surplus that can be sustainably harvested over a defined period of time. The last consideration is generally reserved for game species. Some opponents of rattlesnake roundups argue that because of the annual harvest of the western diamond-backed rattlesnake, this animal certainly qualifies as a game species and should be classified as such in Texas.

The population size of any wildlife species increases or decreases due to two opposing natural forces, called "biotic potential" and "environmental resistance." Biotic potential is the number of offspring that a species is capable of producing under ideal conditions. Some examples of biotic potential include a high reproductive rate and ability to migrate, invade new habitats, and cope with adverse environmental conditions. Environmental resistance is defined as those environmental factors that tend to decrease population numbers. Some examples of environmental resistance include continue if predators remain absent from prey habitat. This example illustrates just one tenuous predator-prey relationship in a network of other such relationships operating in nature's ecosystems.

The purpose of our discussion of the predator-prey balance is to demonstrate the potential impacts of removing predators—of any kind—from their natural habitat. However, even though rattlesnakes and rodents represent a classic predator-prey interrelationship, there are no scientific studies documenting an increase in rodent populations after the removal of the western diamond-backed rattlesnake from its natural habitat. Therefore, to conclude that the hunting of rattlesnakes for round-ups in itself has negative impacts on the predator-prey balance is difficult. Other considerations would include the movement of unhunted rattlesnakes into the habitat where snakes have been removed by hunting; how other rodent predators would prosper and increase by consuming the same prey species as did the hunted rattlesnakes; and whether those rattlesnakes that missed being captured have genetic adaptations to avoid capture and then pass these traits on to their offspring.

lack of food, water, or suitable habitat; and presence of predators, diseases, parasites, and competitors.

It should be pointed out that a single factor (e.g., predators) does not always act as a population regulation/control mechanism. Snakes and other predators represent one environmental resistance factor controlling the high reproductive rates of rodents. If snakes and other predators are removed from natural habitats, then the high biotic potential of rodents, when left unchecked, can cause their numbers to increase quickly and dramatically, leading to a self-inflicted population crash (see later discussion of predator-prey balance). Of note is the fact that the development of human settlement and communities usually includes removing all predator species that could adversely affect human health or economics. This is why mice and other rodents are commonly found living in urban residences and communities, and rattlesnakes are not.

Urbanization is a good example of what happens when environmental resistance factors offset a species' biotic potential. For example, when rattlesnake habitat is destroyed and replaced by parking lots and shopping malls due to urban sprawl, rattlesnakes have three choices: adapt, move, or die. There are many publications in the professional scientific literature attesting to dramatic losses of snakes and other reptiles due to urbanization processes (e.g., road building, habitat fragmentation or loss, and introduction of exotic species).

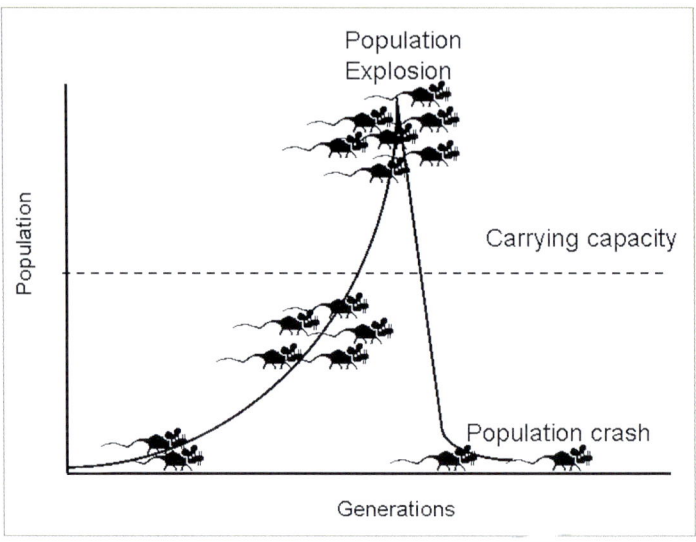

Figure 5. *Diagram of a rat population explosion which could result from the lack of natural predators upsetting the predator-prey balance (by Clark E. Adams).*

Hibernation

For the western diamond-backed rattlesnake, winter sleep (hibernation) begins in November in northwest Texas. It is estimated that 55 percent of rattlesnakes survive overwintering. The survival rate is higher in adult rattlers, particularly those that live in the warmer climates of Texas. Rock crevices, mammal holes, washouts, or caves often serve as hibernation sites (called hibernacula or dens). Snakes disperse only one to two miles from these dens during the spring and summer. One study of a male radiotagged rattlesnake documented that it hibernated under the same rock during three winter seasons.

Young rattlesnakes leave their mothers within a few days after birth, but when it is time to hibernate in the winter, they follow their mother's scent trail and use her den. Subsequent generations also use the same den—some lineages using the same den for decades. Whether individual rattlesnakes in North Texas hibernate in the same dens year after year is unknown, although snake hunters do report capturing rattlesnakes in the same dens year after year.

Reproduction

Western diamond-backed rattlesnakes mate once a year after their emergence from hibernation. In Texas, mating can occur as early as January and usually peaks in March and April. Female rattlesnakes in northwest Texas become sexually mature in three years, and probably produce offspring once every two years, but some may reproduce annually in warmer areas where the growing season is longer.

When male rattlesnakes participate in battles for females, each raises the upper part of its body above the ground to buffet and intertwine with the other (see fig. 6). Common among pit vipers, this "combat dance" of the western diamond-backed rattlesnake is thought to be not only a contest for the right to mate but also a territorial ritual that is not restricted to any one season. Both sexes of the western diamond-backed rattlesnake have anal musk glands in the tail that are used for mate recognition and as a warning device by the snake if it is handled or mistreated. The tail of the male is longer than that of the female. The male has a double-knobbed copulatory organ called the hemipenes, which anchors the male to the female during mating.

Western diamond-backed rattlesnake births occur mainly in late summer or early fall, with an average of ten (range = 4 to 23) young produced per female. One commercial rattlesnake dealer said he extracted forty embryos from one snake. Rattlesnake embryos are retained and develop within the body of the female—the process is known as "ovoviviparous" egg development. The young rattlesnakes develop

Figure 6. Combat dance of male western diamond-backed rattlesnakes to determine mating dominance (by Clark F. Adams).

within gelatinous eggs that remain inside the female's oviduct (a sac that contains the developing embryos) until the time they are ready to emerge. The gestation period (time for embryos to develop into baby snakes) lasts 167 days. The birthing process may last for three to five hours. The young are born encased in a thin egg membrane. They puncture the membrane, crawl out, and immediately begin the survival process like any adult rattlesnake. Maternal care of newborn western diamond-backed rattlesnakes is limited or nonexistent. The rattlesnake birthing process is mistakenly portrayed by dealers at roundups as a baby rattler actually hatching from a hard-shelled egg (see fig. 7).

Danger to Humans and Livestock

Venomous snake bite is a relatively uncommon event in the United States. According to the Texas Department of State Health Services, about 7,000 people are bitten by venomous snakes in the United States annually. This figure may seem large, but when considered in terms of our population of more than 300 million, the problem is put in a more realistic context. Only 0.2 percent (1 out of 500) venomous snake bites result in death. Roughly half of all venomous snake bites are "dry," and on average one to two people in Texas die each year from venomous snake bites (species not reported).

On the other hand, dog bites are

Figure 7. Typical trinket sold at rattlesnake roundups, giving the false impression that baby western diamond-backed rattlesnakes hatch from eggs (by Jessica Alderson).

relatively common events in the United States. Approximately 20 people die each year in the United States as a result of dog attacks. Children are more likely to be bitten than any other segment of the population. In Texas, more than 40 percent of severe bite victims are children under age eleven. More than half of all children will be victims of a dog bite by the time they reach age twelve. Annually an estimated 5 million dog bites occur nationwide, with approximately 400,000 bites occurring in Texas. Dog bites cost insurance companies $250 million a year in medical and liability costs, accounting for one third of all liability claims against homeowner policies.

Even though spider bites are probably more common than snake bites, most cases are minor and statistics are not available globally. Texas has two venomous spiders, the black widow *(Latrodectus mactans)* and the brown recluse *(Loxosceles reclusa)*. The black widow's venom is reportedly fifteen times more toxic than the venom of the prairie rattlesnake.

Nevertheless, people who repeatedly handle venomous snakes are going to be bitten eventually—one can only defy the laws of probability for so long. An experienced snake handler at the Sweetwater roundup claimed to have been bitten forty-two times, but only one bite caused tissue damage. A different snake handler,

who had been collecting live rattlesnakes since he was seven years old, claimed to have been bitten thirty-three times. He and other handlers recounted the intense pain associated with snake bites that had to be treated.

Even though the incidence of rattlesnakes biting people is fairly low, we found several bizarre cases in published articles. One incident occurred when a young boy was bitten on the foot by the severed head of a rattlesnake that still injected venom. In another case, a long-time snake handler and taxidermist was bitten by a rattlesnake that was thawing out after being in a freezer for sixteen hours. The amount of venom received during the bite required surgical removal of muscle tissue on his forearm. There was even an account of venom transfer to a human from a preserved rattlesnake head; freeze drying is a pretty drastic form of preservation, but this incident required antivenin treatment at a hospital.

Sometimes limbs and/or fingers can be lost due to the tissue destruction caused by rattlesnake venom. For example, one hunter who brought a thousand pounds of rattlesnakes to a 1991 roundup arrived in leather sandals, cut-off shorts, and a tank top shirt, and he lacked one arm. The loss of his arm was due to a rattlesnake bite that had not been treated properly (see First Aid for Snake

Bites sidebar, later in this chapter). Nevertheless, this did not deter his continued efforts to hunt rattlesnakes for roundups.

Unlike in the preceding instance, some snake bites may cause a person to consider a career change. This was the conclusion of two Colorado men who routinely worked with rattlesnakes and had been seriously bitten. One was bitten while weighing a bagged rattlesnake. The snake lunged and bit him through the bag. The other victim nearly died from a bite on his thumb.

Occasionally, rattlesnake bites can be lethal. There was one account of an individual who died from a bite to the thumb during a "sacking contest" (contests are discussed in chapter 3). Members of some religious sects, such as the Holiness Movement, dance with venomous snakes to prove their faith, as described in two passages in the New King James Version of the Bible. One passage is found in the book of Mark, chapter 16, verse 18: "They will take up the serpents; and if they drink anything deadly, it will by no means hurt them; they will lay hands on the sick and they will recover." The other is in Luke 10:19: "Behold I give you the authority to trample on serpents and scorpions, and over all the power of the enemy, and nothing shall by any means hurt you." During a church service in November 2006, a forty-eight-year-old

woman was bitten. She died three and one-half hours later. Except for a few brief newspaper accounts, there were no published details concerning the events that caused the lethal bite. It is unknown whether she was dancing with a rattlesnake to prove her faith or was an unfortunate bystander who was in the wrong place at the wrong time.

During the 2006 International Champion Rattlesnake Races held in Brownwood, Texas (see figure 21), "a seasoned veteran snake handler was bitten pretty bad because he put his hand in front of a snake while trying to haze it down the lane. He was in the hospital for several days and was administered several vials of antivenin" (Jackie Bibby, pers. comm.).

Finally, we found an account of a snakebite victim on a golf course. The golfer was reaching down to retrieve his ball from a marshy area when he felt something scratch him on his right temple. He saw a large rattlesnake, passed out, and awoke three days later in the hospital. He had been bitten twice on the face.

Outdoor recreationists who venture into rattlesnake habitat should exercise certain precautions to prevent detrimental encounters with venomous snakes, and many do so (see How to Avoid Rattlesnakes). But others may be at risk even on the most casual foray. For example, a spring tradition in Texas involves

Figure 8. Rattlesnake in a bluebonnet patch (by Nicholas Cowey).

HOW TO AVOID RATTLESNAKES

1. Walk or hike in areas where the ground is clear, so that you can see where you step or reach with your hands.

2. Use a walking stick to rustle the shrubs along the side of the trail to alert snakes and other animals of your presence.

3. Wear protective clothing such as long heavy pants and high boots.

4. Wear gloves when using your hands to move rocks or brush.

5. Watch where you step, and never put your hands in areas where you cannot see.

6. Avoid ledges, cracks, or holes, which are common areas where rattlesnakes can be found resting.

Figure 9. Children in a bluebonnet patch (by Kira Shelby).

thousands of pictures being taken of children and adults in the beautiful bluebonnet patches that grow along the state's highways. This is done without considering the possible presence of western diamond-backed rattlesnakes that might also be in the bluebonnet patch (figs. 8 and 9). A few simple and sensible procedures to follow prior to a bluebonnet photo shoot will make it much safer.

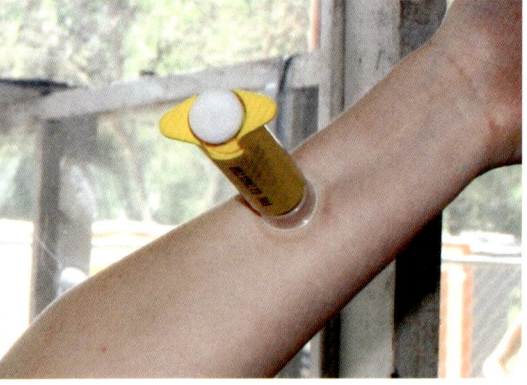

Figure 10. Suction syringe treatment for snake bite (by Clark E. Adams).

During our investigations of rattlesnake roundups, we witnessed two rattlesnake bite incidents. One was at the roundup in Taylor, Texas, where a "sacker" contestant was bitten on the hand. In this case, not enough venom was injected into the puncture wound to cause problems or require hospital treatment. A suction syringe (fig. 10) was applied to the bite, and the sacker went back to the sacking contest that afternoon.

In the second incident, we observed a snake bite to another experienced handler at the roundup in Freer, Texas. This handler was moving five- and six-foot-long rattlesnakes from a commercial dealer's supply box and putting them in a display pit. As shown in figure 11, the handler did not seem to have full control of the snake. He supported and restrained the agitated snake's body and head poorly as he relocated the animal. Before we could say, "He's going to get bit," it happened. The snake whipped its body around to the handler's left forearm and delivered a full load of venom. The handler attempted to shake the snake off his arm. This action caused the snake's fangs (nearly two inches apart) to sink even deeper into its victim's forearm. In panicked desperation, the bitten handler was able to separate himself from the rattlesnake. He ran to the rear of the tent where he passed out and collapsed fifteen feet from the exit.

Figure 11. Snake handler who, minutes after this picture was taken, was bitten and suffered the consequences of not securing the head of a very large western diamond-backed rattlesnake (by John K. Thomas).

Organizers and workers attempted suction treatment with syringes (fig. 10) and even tried to electroshock the victim under the mistaken notion that an electric current would crystallize the venom and render it impotent. Common sense eventually prevailed; those trying to help the victim transferred him to a hospital in Alice, Texas, thirty-six miles away. Six months later, we learned from the victim that he had recovered after extensive treatment involving several antivenin injections, a fasciotomy (i.e., removal of dead muscle tissue in his forearm), and six months of recuperation.

The American Red Cross provides a comprehensive overview of appropriate actions that need to be taken in the event of a venomous snake bite (see First Aid for Snake Bites).

There is little historical evidence of rattlesnakes being a serious threat to livestock and pets. In 1950, Texas county extension agents were given a survey to assess the seriousness of damage caused by rattlesnakes. Eighty-two percent of the 134 responding agents rated damage from rattlesnake bites to livestock and pets as negligible or unimportant. Cattle were the most frequently bitten, compared to horses, sheep, goats, and pigs. We asked Dr. Bud Alldredge, a veterinarian who practiced in Sweetwater, Texas, in 1991, about livestock losses due to rattlesnake bites. Although he agreed that most snake bites to livestock produce negligible affects, he had observed losses of dogs, cats, lambs, and other young livestock due to rattlesnake bites in the area where he practiced. His first patient, when he established his practice in Sweetwater, was a horse that was bitten by a rattlesnake; the horse survived. The next snake-bitten horse that he treated died.

It would seem that if the reason for dispatching thousands of rattlesnakes during roundups involves a real or perceived danger to people and livestock, there should be some sort of procedure in place to document this danger on an annual basis. It has been more than fifty years since anyone has attempted to examine the magnitude of negative impacts on human health and economics due to rattlesnake bites.

Myths about Rattlesnakes

A myth is an account or story that people believe about something around them—in this case, rattlesnakes. Many myths are based on old folk tales and misinformation fueled by fear, humor, or rumor. Some of the more common myths about rattlesnakes are listed below. Others can be found in Jack Kilmon and Hooper Shelton's book *Rattlesnakes in America*.

1. *Rattlesnakes can grab on to their tails to form wheels or hoops that allow them to roll down hills at speeds faster than they can achieve by crawling on their bellies.* There is no evidence that this is true or even anatomically possible for any snake to do.

2. *Rattlesnakes are not aggressive and just want to be left alone.* Believe it or not, rattlesnakes have different temperaments. Some have to be provoked intensely before they sound a warning or strike in defense. Others have nasty dispositions and will respond to the slightest disturbance. Snake handlers try to identify the more aggressive rattlesnakes and avoid using these in their roundup shows. This was the case in the 2006 Taylor roundup. A particularly agitated rattlesnake was among those in the exhibition ring with the other snakes. The snake handlers noticed it immediately and proceeded to take actions to remove it from the ring. What ensued was a draconian effort on the part of three very large men (easily 200 or more pounds each) to subdue, restrain, and bag a 1.5-pound rattlesnake. Even though the snake was removed from the ring, it still had the parting shots, spraying the snake handlers with pheromones from its anal gland and with venom from its fangs.

This appeared to be a classic example of a rattlesnake with an "attitude," one that should have been left alone.

3. *Rattlesnakes always sound a warning with their rattles before striking.* A rattle is a warning mechanism that a rattlesnake uses when threatened. It would not use its rattle before striking prey but would use it to warn a hiker who had inadvertently surprised it.

4. *Rattlesnakes can strike only from a coiled position.* A rattlesnake can strike from any position. None of the western diamondbacked rattlesnakes that bit handlers at the roundups we attended were coiled before their strikes.

5. *Rattlesnakes can strike at a distance twice their body lengths.* Most rattlesnakes can strike at a distance about one-half their body length.

6. *Removal of a snake's fangs makes it harmless.* Rattlesnakes lose and break fangs all the time. They are able to replace a fang from the seven sets of immature fangs in the rear of their mouths within a day or two.

7. *Rattlesnakes add one rattle per year of age.* Rattles are added when a snake sheds its skin, which can occur up to six times a year, depending on how fast it grows.

8. *A baby rattlesnake is not as venomous as an adult snake.*

FIRST AID FOR SNAKE BITES

Symptoms of human envenomation include: pain and gradually increased swelling around the bite site; a metallic taste; difficulty breathing; chest pain; nausea, vomiting, or diarrhea; weakness and burning or prickling sensations; and dizziness and fainting.

According to the American Red Cross, a person should take the following steps when bitten by a snake:

- Wash the bite with clean water and soap.
- Immobilize the bitten area and keep it lower than the heart.
- If the bite is on the hand or arm, remove any rings, watches or tight clothing.
- Get medical help immediately.

Most bites do not occur in isolated situations where the victim may be a long distance from medical help. If this is the case, medical professionals, including the American Red Cross, cautiously recommend two other measures. If a victim is unable to reach medical care within thirty minutes, he or she should apply a bandage, firmly wrapped two to four inches above the bite, to slow the venom's movement through the bloodstream. The bandage should limit, not cut off, blood flow in a vein or artery. A good rule of thumb is to make the bandage loose enough that a finger can be slipped under it. The Red Cross also suggests that a suction device may be placed over the bite to help draw venom out of the wound. Suction instruments (fig. 10) often are included in commercial snakebite kits.

U.S. medical professionals do not always agree about first aid methods to treat a snake bite, but they are nearly unanimous in their views of what *not* to do:

- Do not apply ice or any other type of cooling on the bite, since this can be harmful.
- Do not apply tourniquets that cut off blood flow completely and may result in loss of the affected limb.

- Do not use electric shocking, which has yet to be proven effective and can harm the victim.
- Do not make incisions in the puncture wounds, which, improperly done, could cause more infection and possibly damage to an artery or vein, resulting in great blood loss by the victim.

The administration of antivenin in a clinical setting is the safest treatment for a rattlesnake bite. Antivenin is produced by first "milking" the venom from a rattlesnake (see chapter 3). An antivenin manufacturer then dilutes the venom and processes it for later injection into a horse, goat, or sheep. After the animal is administered several increased dosages, it builds up immunity by developing antibodies or blood proteins that protect the animal from the venom's toxic effects. These antibodies accumulate in the blood and are extracted from the animal as a serum that is centrifuged to separate out the red blood cells. This serum is then purified and marketed to medical facilities to treat human victims.

The package insert for the antivenin Crotalidae Polyvalent Immune Fab (Ovine), for example, provides the following detailed information about its range of effect on various venomous snake bites:

CroFab® is a sterile, nonpyrogenic, purified, lyophilized preparation of ovine Fab (monovalent) immunoglobulin fragments obtained from the blood of healthy sheep flocks immunized with one of the following North American snake venoms: *Crotalus atrox* (Western Diamond-backed rattlesnake), *Crotalus adamanteus* (Eastern Diamond-backed rattlesnake), *Crotalus scutulatus* (Mojave rattlesnake), and *Agkistrodon piscivorus* (Cottonmouth or Water Moccasin). To obtain the final antivenin product, the four different monospecific antivenins are mixed. CroFab is standardized by its ability to neutralize the lethal action of each of the four venom immunogens following intravenous injection in mice.

Source: http://www.protherics.com/

Rattlesnakes are born with fangs, venom, and the ability to strike and bite. They pose the same level of threat to people as an adult rattlesnake.

At this point in our discussion we invite you to test your own knowledge about rattlesnakes and their interactions with humans. A list of statements appears in the sidebar Fact or Fiction? The answers to several of the statements have been revealed in earlier sections of the book. The correct response for each statement is given at the end of the list.

FACT OR FICTION?

Are the following statements (a) true, (b) false, or (c) idle speculation?

1. More rattlesnakes are killed on roads by automobiles than by hunters for rattlesnake roundups.
2. Rattlesnakes can control the amount of venom they discharge through their fangs during a strike.
3. The most dangerous rattlesnake is one that has lost all of its tail rattles.
4. Rattlesnake babies hatch from eggs.
5. More people die from wasp stings than rattlesnake bites.
6. If you are bitten by a venomous snake, you are much more likely to die in an auto accident on your way to the hospital than from the snakebite itself!
7. People can die from nonvenomous snake bites as well as bites from venomous snakes.
8. Rattlesnakes have the ability to hypnotize or charm their prey so they can't flee.
9. People in cities or suburbs who receive a rattlesnake bite should go to the hospital closest to the zoo.
10. Camouflage clothing is necessary attire during a rattlesnake hunt.
11. Female rattlesnakes will swallow their young to protect them from danger. Once out of danger, the female spits them out.
12. Rattlesnakes cannot close their eyes when they sleep.
13. A rattlesnake is still venomous if you remove its fangs.
14. Livestock losses due to rattlesnake bites are a major problem.
15. A rattlesnake injects venom with every bite.

Answers: 1 = c, 2 = b, 3 = a, 4 = b, 5 = a, 6 = a, 7 = a, 8 = b, 9 = a, 10 = b, 11 = b, 12 = a, 13 = a, 14 = b, 15 = b.

Chapter 2

History of Texas Rattlesnake Roundups

Although many fear snakes, people are also curious about them and want to know more about them. Human curiosity is a major reason why people attend and participate in rattlesnake roundups. These roundups have traditionally been held in rural communities located in Alabama, Florida, Georgia, Kansas, New Mexico, Oklahoma, Pennsylvania, and Texas. Outside Texas, the town of Okeene, Oklahoma, claims the oldest roundup, dating back to 1939, although evidence exists that paid (bounty) hunts were conducted as early as 1680 in Massachusetts and 1849 in Iowa. In Texas the Sweetwater rattlesnake roundup has the undisputed reputation as the largest active roundup in terms of public attendance and number of rattlesnakes brought in by hunters.

Rattlesnake roundups were historically one of the first organized methods to address real and perceived threats of rattlesnakes killing or injuring people and livestock. Roundups started primarily for the purpose of reducing the number of rattlesnakes in the area. Hunters ceremoniously captured and killed thousands of rattlesnakes and buried them in mass graves. Some of the meat was fried and eaten, and rattles were saved as souvenirs. Sometimes roundups included rattlesnake shoots (fig. 12) in which contestants could test their marksmanship on a live

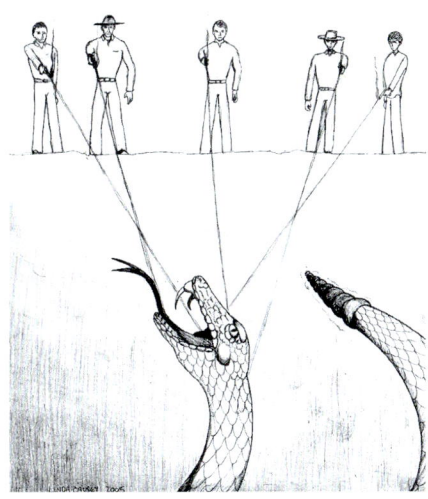

Figure 12. What a rattlesnake shoot might have looked like (by Linda Causey).

rattlesnake. Such shoots were conducted in Texas as late as 1989. The city of Clairemont in Kent County, now a ghost town, was the last Texas community to sponsor rattlesnake shoots (see Annual Peace Officers Rattlesnake Shoot).

In addition to shoots, roundups historically included rattlesnake decapitation and stomping contests. Decapitation contests were timed events in which the winner decapitated the most live rattlesnakes during an allotted period of time. Stomping contests consisted of putting a live rattlesnake in a potato sack and recording how many boot stomps were required to kill the snake. In another form of entertainment, seemingly bold spectators draped a live rattlesnake around their neck, but only after each snake's mouth had been sewn shut with a steel leader. Overall, the entertainment at early rattlesnake roundups focused on methods of dispatching an animal despised by the public and considered to be a useless and dangerous aspect of their environment. Although captured rattlesnakes were used to generate several activities for public entertainment, they all suffered the same outcome.

ANNUAL PEACE OFFICERS RATTLESNAKE SHOOT

CLAIREMONT(Special)—The 41st Annual Peace Officers meeting and Rattlesnake Rodeo will be held today in the Clairemont 4-H Barn.

This year's event is dedicated to Johnnie Holbrooks, Lake Ranger for White River Municipal Water District in southeast Crosby County.

As in the past, the highlight of the rodeo will be the Rattlesnake Shoot where peace officers, as well as visitors, compete for points by shooting at live snakes. Awards will be presented in five categories: masters division, trophy and $150 pair of boots; first place, trophy and $150 pair of boots; second place, trophy and $100 Bailey hat; third place, trophy and $50 gift certificate; and guest and visitors first place, trophy. Shooters get 10 points for hitting the snake in the head or two inches of the nose and 5 points for hitting them anywhere else in the body.

Other activities include a horse shoe pitching contest and riot gun trap shooting.

Registration will begin at 9 A.M. today. Trap shooting and horse shoe pitching will begin at 9 A.M. and continue until all contestants have finished.

A catered barbeque lunch will be served at noon. The rattlesnake rodeo will begin at 2 P.M. Awards will be presented at 3 P.M.

Source: Lubbock Avalanche-Journal, *May 11, 1989.*

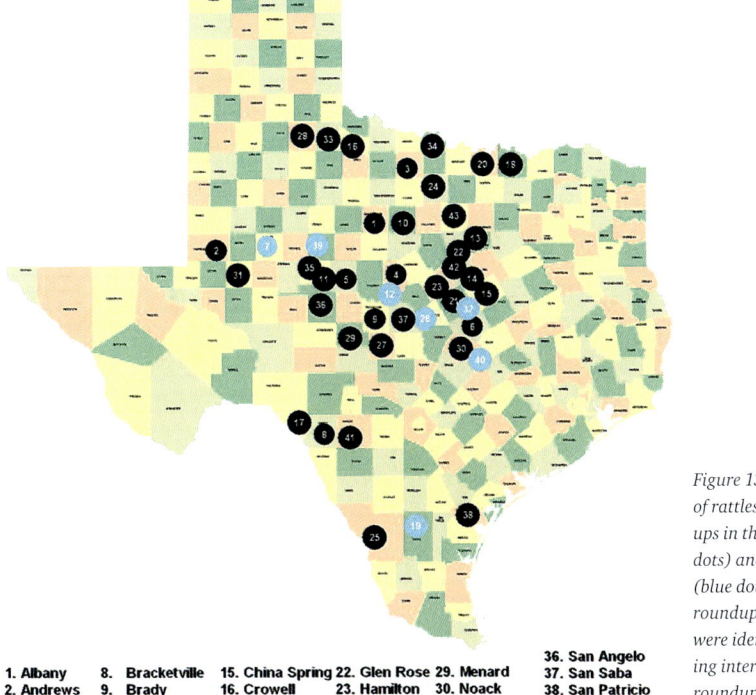

Figure 13. Locations of rattlesnake roundups in the past (black dots) and present (blue dots). Past roundup locations were identified during interviews with roundup hunters, dealers, and organizers (by Clark E. Adams).

1. Albany	8. Bracketville	15. China Spring	22. Glen Rose	29. Menard	36. San Angelo
2. Andrews	9. Brady	16. Crowell	23. Hamilton	30. Noack	37. San Saba
3. Archer City	10. Breckenridge	17. Del Rio	24. Jacksboro	31. Odessa	38. San Patricio
4. Bangs	11. Bronte	18. Elmont	25. Laredo	32. Ogelsby	39. Sweetwater
5. Ballinger	12. Brownwood	19. Freer	26. Lometa	33. Paducah	40. Taylor
6. Belton	13. Cleburne	20. Gainsville	27. Mason	34. Petrolia	41. Uvalde
7. Big Spring	14. Clifton	21. Gatesville	28. Matador	35. Robert Lee	42. Walnut Springs
					43. Weatherford

Geographic Distribution of Past and Present Rattlesnake Roundups

Based on personal interviews with rattlesnake hunters, dealers, handlers, and roundup organizers and reading old newspaper accounts, we learned the locations of past and present rattlesnake roundups in Texas (fig. 13). The locations of roundups and the territorial range of the western diamond-backed rattlesnake largely coincide. The geographical distribution of past and present roundup locations in Texas is within the known range of the western diamond-backed rattlesnake, mainly west of the Interstate Highway 45 corridor (fig. 14). Four of the seven rattlesnake roundups that were still active in 2006 are in communities situated along major state and interstate highways and close to major population centers in the western diamond-back's range.

Western diamond-backed rattlesnakes comprise the majority of snakes brought to roundups, but we observed several species of non-venomous snakes being sold at the

Freer roundup in 1991. The western diamond-backed rattlesnake is found throughout much of Texas, except in a few counties east of the Interstate 45 corridor (fig. 14). This species' range includes 111 million acres in at least 172 of the 254 counties in the state. However, rattlesnake roundups occur in only seven counties within the statewide range of the western diamond-backed rattlesnake. This observation suggests that rattlesnake hunting for roundups could have a limited regional influence on the statewide population of rattlesnakes. This limited regional influence was corroborated when the counties where hunters actually harvested western diamond-backed rattlesnakes for roundups was determined. This information is presented in the sections Rattlesnake Hunters and Rattlesnake Hunts in chapter 3.

Changes in Community Locations of Rattlesnake Roundups

More rattlesnake roundups have come and gone than are presently operating in Texas. Prior to 1991, forty-three Texas communities held rattlesnake roundups. One of the earliest records

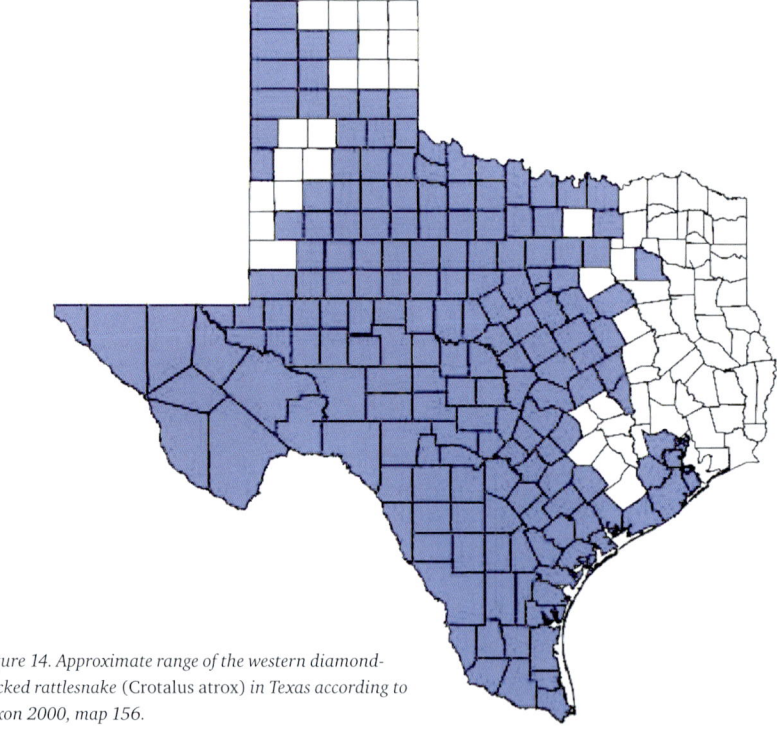

Figure 14. Approximate range of the western diamond-backed rattlesnake (Crotalus atrox) *in Texas according to Dixon 2000, map 156.*

of a community rattlesnake roundup was in Paducah (Cottle County) in the 1920s. Figure 13 shows the community locations of past roundups and the seven remaining active rattlesnake roundups in Texas in 2006.

Possible reasons why several rattlesnake roundups were discontinued include a decline in public attendance, lack of organizational sponsorship, and too much competition with other roundups. Other factors with bearing are the circumstances of costs to hunt snakes exceeding their market value, a decline in restaurant and public demand for rattlesnake meat and other related products, and inconvenient locations away from major highways. New Texas Parks and Wildlife Department (TPWD) regulations designed to monitor rattlesnake harvests probably had an effect, as did community inability to transfer the roundup tradition from older to younger generations of residents.

The decline in the number of rattlesnake roundups may also have been affected by increased concern about animal cruelty, as expressed by environmental and animal rights groups, the Texas public, and the nation. The Humane Society of the United States charges that "rattlesnake roundups are cruel and ecologically damaging events that have taken place in many parts of the United States since the late 1920s. Every year, in the states of Texas, Oklahoma, Kansas, New Mexico, Pennsylvania, Alabama, and Georgia, thousands of rattlesnakes are captured and slaughtered or used in competitive events in ways that violate the most basic principles of wildlife management and humane living. Though they are often promoted as fundraising events for local civic causes, rattlesnake roundups primarily benefit their organizers and corporate sponsors" (The Truth Behind Rattlesnake Roundups, http://www.hsus.org/ace/12078).

POSITION OF THE AMERICAN SOCIETY OF ICHTHYOLOGISTS AND HERPETOLOGISTS CONCERNING RATTLESNAKE CONSERVATION AND ROUNDUPS (2006)

The American Society of Ichthyologists and Herpetologists, an international society of about two thousand professional scientists who specialize in the biology and conservation of fishes, amphibians, and reptiles, strongly opposes traditional rattlesnake roundups. Such roundups promote overexploitation of natural populations of wildlife, unnecessary killing and inhumane treatment of individual animals, degradation of habitat, and promotion of outdated attitudes toward important elements of America's natural heritage. Found nowhere but in the Americas, and especially diverse in the United States, the more than thirty species of rattlesnakes comprise a distinctive component of North America's biodiversity, and one that is increasingly imperiled.

Additionally, wildlife biologists, ichthyologists (who study fish), herpetologists (who study amphibians and reptiles), and other scientists have voiced their concern about the methods used to capture and exterminate rattlesnakes (see position statement sidebar). Certain laws and U.S. Fish and Wildlife Service regulations became important considerations, including the Endangered Species Act, which forbids the capture of threatened and endangered species of rattlesnakes, such as the timber (canebrake) rattlesnake (*Crotalus horridus*), for roundups. Finally, the size of Texas' urban population has increased substantially while the rural populace has declined since the 1950s. Urbanization and urban sprawl resulted in loss of the rattlesnakes' natural habitat and reduced the number and size of rural communities that could economically support roundup events.

The number of years that communities have held a rattlesnake roundup varies from a few years to several decades. Table 1 gives some examples of how long some communities have conducted rattlesnake roundups and, in some cases, when they were discontinued.

Some of these roundups were started with the help of organizers of others. For example, the original organizers of the Sweetwater roundup sent a delegation to the longstanding roundup in Okeene, Oklahoma, in 1958 to gather ideas on how to plan and operate their own event. In turn, the Sweetwater Jaycees helped the Texas Lions Club in China Springs start a rattlesnake roundup in 1989. The organizers in Taylor adopted the roundup in Noack. Ten roundups were discontinued between 1988 and 2006 for reasons already identified. El Paso and Lubbock organizers experimented with rattlesnake roundups in the mid-1990s but subsequently dropped their efforts.

San Patricio is not included in the list because it conducts rattlesnake races rather than a roundup. A rattlesnake race consists of coaxing a rattlesnake down a track by blowing air behind the snake through a plastic tube. This procedure is discussed later and illustrated in figure 21. The San Patricio event has existed for the past thirty-three years, but the organizers have not bought snakes since 1979. They lease snakes from commercial dealers to be used in the races. Nevertheless, attendance at the San Patricio event swells the town's population of 245 to nearly twenty-five times its normal size. The income derived from the six thousand participants is used for the restoration of historical buildings and other community projects.

More than a hundred thousand people attended the sixteen roundups held in 1991. It is instructive to compare roundup attendances to

TABLE 1. HISTORY OF RATTLESNAKE ROUNDUP PARTICIPATION IN SELECTED TEXAS COMMUNITIES UNTIL 2006

Community	Number of years a roundup was conducted	Year roundup was discontinued
Andrews	6	1990
Archer City	27	1996
Big Spring	35	Still active
Breckenridge	10	1995
Brownwood	31	Still active
China Spring	3	2000
Cleburne	15	2000
Freer	40	Still active
Gainsville	33	2001
Jacksboro	36	1991
Lometa	35	Still active
Oglesby	29	Still active
San Angelo	7	1988
Sweetwater	47	Still active
Taylor	34	Still active
Walnut Springs	37	2003
Weatherford	12	1991

TABLE 2. CHANGES IN COMMUNITY POPULATION SIZE BEFORE AND DURING RATTLESNAKE ROUNDUPS

Community	Before Roundup	During Roundup
Freer, Texas	3,418	19,588
Sweetwater, Texas	12,208	42,500
Walnut Springs, Texas	661	7,050

note how much the population size of relatively small communities increased during a roundup weekend (see table 2). Attendance information may be more of an estimate than actual numbers. Many roundup organizers did not take attendance because it was too much trouble, and they did not know what to do with the information.

The crush of people going through the pavilion gate at the 2006 Sweetwater rattlesnake roundup was amazing. The organizers could hardly keep up with the hand-to-hand money exchange with paying spectators, much less count them. Although attendance cannot be precisely quantified in every case, these communities probably experienced fairly significant increases in outside revenue for the two to three days of roundup activities. For example, assume that thirty thousand people attended the 2006 Sweetwater rattlesnake roundup. Based on parties each consisting of two parents with one child and ticket prices of $6 for adults and $3 for children, the community would gross $150,000 (20,000 x $6 + 10,000 x $3) just to let people into the arena. Sweetwater estimated in a study that over a three-day period, its roundup produced a $3 million-dollar impact for the community.

Chapter 3

Organization of Texas Rattlesnake Roundups

Texas rattlesnake roundups have been organized around three key components: (1) the availability of a rattlesnake resource, which attracts (2) commercial dealers, and which leads to (3) a community group serving as organizers of a rattlesnake roundup. Commercial dealers and organizers operate with different agendas, and each has a different level of dependence on the rattlesnake as a resource. The organizers plan the roundup around the specific needs and anticipated contributions of the hunters, snake handlers, vendors, and spectators. Outside the formal organization of the roundup are the TPWD, protest groups, and other commercial dealers who have vested interests in rattlesnakes but are isolated from the decisions involved in the organization of or proceedings at rattlesnake roundups.

Commercial Dealers

There are two types of commercial dealers: the dealer who has a con-tract with the roundup organizer and the dealer who does not. Both types of dealers provide hunters opportunities to sell and dispose of captured rattlesnakes. Both types also produce a measurable impact on roundup proceedings. During the 1970s, commercial dealers established markets for rattlesnakes. They paid twenty-five cents per pound of live rattlesnake. The price doubled in 1975 to fifty cents and quadrupled to $2.00 per pound in 1977. In 1981, it doubled again to $4.00 a pound. Live rattlesnakes sold for a high of $8.50 per pound in 1989. The high price per pound led some hunters to use drastic measures, such as bringing in backhoes to excavate dens to gain access to rattlesnakes. Prices later declined to $3.50 and $5.50 a pound in 1990 and 1991, respectively. Of course the price per pound was driven by public purchases of the many artifacts made from rattlesnakes' anatomical parts. By 2006, the price paid to hunters had declined to $2.00 per pound. At this time there was a market surplus of rattlesnake hides, so buyers were not paying very much for snakes. By 2007, the price had declined to fifty cents per pound.

Organizers generate additional income by seeking competitive bids and signing a contract with a specific dealer who has exclusive rights to all of the snakes brought to the roundup. The dealer offers to pay a certain price per pound for all of a roundup's rattlesnakes. The organizers negotiate a bid price that is higher than what they pay hunters. The difference between the bid and pay prices is "profit" for the roundup. A contract is the legal agreement by the organizers to pay a particular dealer his bid price for the rattlesnakes brought to their roundup.

Contracted dealers also play other integral and important tasks at roundups. At some roundups, a commercial dealer checks in hunters, pays them for their snakes, measures and weighs snakes entered into the roundup contests, moves snakes from boxes to pits, and removes snakes after the completion of the roundup. Sometimes dealers bring their own snakes to the roundup in order to enhance the presence of rattlesnakes and the public image of a rattlesnake roundup.

It is interesting to note that there was an almost complete change in the cast of individuals who were dealers operating in 2006, compared to 1991. At the time of writing there are four major dealers in Texas: Diamondback, Mike Ivey, Maverick, and Tumbleweed Traders. Two companies, Take-a-Memory Home and Sleepy Leather, are close affiliates who produce many of the rattlesnake products sold by two of the dealers. John Shaddix with Rare Skins Inc., George Wills with Maverick Trading Post, and Bob Popplewell used to be major players in the early 1990s but have since gone out of business, moved elsewhere, begun to market other products, died, or retired. Maverick and Tumbleweed Traders now have new owners.

Some commercial dealers who do not win a roundup bid might "scalp" the harvest of rattlesnakes for roundup before or during the event. Scalping occurs when snake dealers arrive in a community a week or two before the roundup and buy the hunters' rattlesnakes. They offer a slightly higher price per pound than the dealer contracted to buy the rattlesnakes at the roundup. Buying stations are set up at a motel or other location close to the roundup. Some dealers have scalped their own roundup just prior to the event. While roundup organizers are aware of this situation, they do not care because the majority of their income is not based on the sale of snakes to commercial dealers. The overall effect of scalping is that fewer snakes are brought directly to roundups by hunters, the hunters make a little extra money, and they do not have to hold rattlesnakes for long periods waiting for a roundup to start.

The Texas Parks and Wildlife Department

On occasion, the TPWD receives information from roundup organizers and protest groups, but until recently it has not been involved in regulating rattlesnake harvests and associated roundup activities. Yet, according to the Nongame Species Act as amended in 1985, the TPWD is the state agency responsible for the management of all nongame species in Texas. When asked how many rattlesnake roundups were in operation in 1991, none of the commission members, the director of TPWD, or the head of the Wildlife Division knew the answer, except to say there were probably less than thirty (there were seventeen at that time). It is unlikely that they were aware of the number or location of rattlesnake roundups that were still operating in 2006 with the exception of the Sweetwater roundup.

In the early 1990s, there was only one regulation that applied to rattlesnake collection. This regulation required that in order to pick up or collect a rattlesnake in the field, the person must have a valid Texas general hunting license. However, once rattlesnakes had been captured and transported to the roundup, no state laws existed that regulated their sale

to commercial dealers. Even though the commercial value of rattlesnakes translated into a multimillion-dollar industry, there were no regulations on hunting seasons, methods of capture, or bag limits and no licensing system existed for commercial rattlesnake collectors and hide dealers. Despite being a regularly hunted species, the western diamond-backed rattlesnake has never been considered a game animal like deer, quail, or turkey, which receive regulatory oversight.

If the western diamond-backed rattlesnake were declared to be a game species, TPWD wildlife managers would be required to determine what level of hunting pressure would be appropriate to sustain a huntable population over time. As with white-tailed deer, this would require a means of estimating the size of the rattlesnake population, probably at the county level of analysis, in order to set harvest limits. The complexity and difficulty of conducting field studies to determine the impact of hunting on rattlesnake populations is discussed in chapter 4. For this reason alone, and there are many others, it appears highly unlikely that *Crotalus atrox* will ever be declared a game species in Texas.

The TPWD eventually recognized the need to impose some form of regulation on the collection and commercial trade of nongame wildlife (see Events Leading to the Nongame

Permit). In 2003, the Texas Parks and Wildlife Commission promulgated a regulation that required commercial nongame collection permits for anyone who collected, sold, or purchased nongame wildlife (Texas Administrative Code, Chapter 65 Wildlife, Subchapter O, titled Commercial Nongame Permit). The chapter also required that permit holders provide annual records of harvest, sales, and lists of species. Only a select group of species are covered under the chapter. The primary focus of subchapter O is to regulate commercial activity involving nongame wildlife. The chapter does not specifically mention rattlesnake roundups or other specific collecting areas.

Oversight and law enforcement have become problems associated with the regulatory intent of initiating the nongame permit. It is a paper tiger: toothless and ineffective. Although years of records—a permit requirement—on rattlesnake harvests have been sent to the TPWD, the information has never been analyzed in any meaningful way. For example, the information is not being used to monitor trends in commercial use of native wildlife species, as is the intent of the nongame permit. We observed contracted dealers checking for nongame permits when hunters brought their rattlesnakes to the roundups we attended in 2006, but we saw no TPWD game wardens at most roundups, except Sweetwater.

Protest Groups

Protest groups typically consist of a mix of individuals who vary in their agendas, levels of knowledge about

EVENTS LEADING TO THE NONGAME PERMIT

Matt Wagner served as nongame program leader for the TPWD from 1992 to 1998. This is the account he provided in December 2006:

We were aware of extensive commercial collection of nongame wildlife with no oversight by the Department. For example, in September, 1993, a food distributor contacted me about obtaining permits for selling 60,000 pounds of soft shell turtles to Taiwan. When I explained to him that there was no permit required, I found out that the turtles were collected out of the wild in South Texas, fed and sold live at a rate of $2.15 per pound for a total of nearly $130,000!

Upon further investigation, it was discovered that at least 40 species of snakes, two dozen lizards and salamanders, 18 species of turtles, and more than a dozen frogs were being bought and sold. Reptiles, including box turtles, graybanded king snakes, green anole lizards, and numerous other reptiles were being sold as pets. In addition, Department staff members were receiving reports of buyers driving circuits in the state to buy rattlesnakes for the commercial trade in rattlesnake natural history, and methods of voicing their opposition to rattlesnake roundups. Nevertheless, they are all bound by a common desire to see an end to rattlesnake roundups and by the need to take advantage of every media opportunity. Their placards present many different messages to roundup organizers and/or spectators, including:

- For crying sakes stop frying snakes.
- Education not exploitation.
- Change the roundup—keep the snakes—stop the killing.
- *Crotalus atrox*: To be protected not exploited.
- Preserve the balance.
- Most snake bites occur at roundup.
- Snakes Not Steaks.

- Snakes have homes, Leave them the hell alone.
- When the snakes are away, the rats will play.
- Gassing dens is like fishing with dynamite.

Protest group members consist of doctorate-level herpetologists, members of herpetological societies, snake fanciers and hobbyists, moms and pops and kids, blue-haired ladies in Reeboks, and animal rights activists who will do just about anything to shut down a rattlesnake roundup (fig. 15). The rattlesnake roundup organizers are just as determined to put on the show as the protest groups are determined to close it down. We observed several examples of such conflict between rattlesnake roundup promoters and protestors. On one

specialty skins and rattlesnake parts. In 1992, at least two large-scale dealers in live rattlesnakes claimed to operate multiple processing facilities throughout the state. At least twelve commercial buyers of rattlesnakes and rattlesnake parts were identified in 1992 and 1993. In addition to the rattlesnake trade, at least 23 species of mammals were also being collected and sold, including prairie dogs, flying squirrels, bobcats, and even bats.

After a series of reports on the extent of the commercial trade of nongame wildlife was presented by department staff to TPWD commissioners, a commercial Nongame Permit was approved in April, 1998. This permit was required for anyone with 25 or more specimens of nongame wildlife, or any person engaged in buying and selling them. The intent of the permit was to monitor trends in commercial use of native wildlife species, and take further action if necessary to protect vulnerable populations.

Figure 15. Protesters at the Taylor rattlesnake roundup in the early 1990s (by McAllister Dowd).

occasion a young man chained and locked himself to the chain-link fence by the ticket booth at the Taylor roundup. Unable to unlock him from the fence, the organizers threw a tarp over him until he became warm enough to concede the contest. We saw protestors (some dressed in skeleton costumes) who lay down on the road in front of the pavilion hosting the Sweetwater roundup. Roundup personnel patrolling the area on horseback had no problem negotiating their steeds around the prostrate protestors. This spectacle was a bonus form of entertainment for spectators going into the rattlesnake roundup pavilion. In 2006 we observed that when there were no media representatives present at a roundup, there were no protesters either—and vice versa.

Local newspapers usually provide most of the media coverage on a community's rattlesnake roundup. Televised media coverage occurred only at the Sweetwater and Taylor roundups during the 1990s. Because of its stature as a rattlesnake roundup, Sweetwater was also visited by journalists and photographers from the National Geographic Society who were documenting the roundup.

During the 2006 Sweetwater roundup, National Public Radio (NPR) interviewed Rodney Kinsey, a Sweetwater Jaycee member and roundup organizer. This interview drew the ire of many listeners and professional biologists because of the cavalier attitude of the NPR reporter toward the eminent deaths of nearly ten thousand rattlesnakes. The reporter tried to provide entertainment rather than an in-depth analysis of the overall ecological consequences of such a large loss of natural predators from their habitats.

The Organizers

In the movie *Field of Dreams*, the phrase "build it and they will come" drove the main character of the film to construct a baseball diamond in his cornfield. At the end of the movie, fans were drawn from far and wide to see a mystical baseball game. Their one-dollar admission fee produced enough revenue to save the farm. Rural communities took a similar build-it-and-they-will-come approach to organizing rattlesnake roundups in Texas. Roundups became competitive as communities vied with one another for spectators and rattlesnakes.

Organizers of past rattlesnake roundups were groups of local citizens who had a general disdain for rattlesnakes and viewed them as a menace and nuisance. They arranged roundups to include hunting, check-in and handling rattlesnakes at the roundup, festive entertainment, and disposition of the snakes after the festivities. Public gratitude was the only compensation they received for their efforts.

Beginning in the late 1950s civic groups such as the Sweetwater Jay-cees began to organize rattlesnake roundups. They were followed in other communities by groups including the Lions Club, Chamber of Commerce, Volunteer Fire Department, Kiwanis, American Business Clubs, and local sportsmen or snake handling clubs. In some communities, when one civic organization discontinued its involvement in a roundup, another would adopt the project. As with the oldest roundups, the primary purpose was to reduce the number of rattlesnakes in the area. Captured rattlesnakes were given to individuals and research groups for a variety of personal uses, or killed and buried in mass graves, since they had little economic value at that time.

More contemporary planners are members of civic organizations who want to produce community income and provide money for charities; they view their participation in roundups as a matter of civic responsibility. Consequently, the focus of roundup programs has broadened during the last quarter century beyond animal control. Organizers promote public education, safety, and entertainment demonstrations. In fact, depending on programmatic emphases, roundups are similar in nature to events ranging from a church social to a demolition derby. Civic groups, which organize these roundups, generate income from gate receipts, the sale of fried rattlesnake meat, and vendor products. The majority of roundup income supports local community projects, and the events provide a chance to socialize and develop a sense of community cohesiveness and civic pride.

Roundup Facilities

Roundup facilities are typically located at the community fairgrounds, where other events such as carnivals, arts and crafts shows, and gun and coin shows are conducted concurrently. Carnivals and specialty shows greatly increase the entertainment experience and appeal to more tourists. Large numbers of tourists and spectators result in more money spent in a roundup community.

A roundup facility generally has seven prescribed areas. One area is for hunters to check in their snakes,

register them for particular competitions, and sell them to a dealer who has the contract for that roundup. The check-in area is often restricted from public access for safety reasons and is usually located away from areas heavily traveled by the public.

A second area is the holding pit (fig. 16) where the rattlesnakes are kept during the roundup. The public is allowed to view the rattlesnakes in this area. A holding/viewing pit is usually twenty feet square with four-foot-high walls built of Plexiglas, plywood, or concrete. It occasionally has a railing two feet away from the walls as a second line of protection for the

Figure 16. Dumping snakes at the Sweetwater snake holding area in 2006 (by Clark E. Adams).

public. A pit can hold several hundred snakes with one or two snake handlers managing the location and movement of snakes. Rattlesnakes bite one another when placed into these rather surreal conditions that do not allow them the freedom of escape or defense.

At the 2006 Sweetwater roundup, handlers dumped hundreds of snakes at one time into the holding pen from as high as four feet. They landed on a concrete floor or on the backs of other snakes. The white bellies of dead snakes in the holding pen were conspicuous. This treatment might have been due to greater handling risks posed by the larger snakes that were collected, or handlers losing their appreciation for the snakes' welfare after handling thousands of snakes over the years.

The third area is for venom extraction and butchering of rattlesnakes for skins, meat, and entrails. Only the Sweetwater roundup butchers snakes for public entertainment and perhaps education. The Sweetwater roundup has an elaborate meat processing area. It is efficiently organized to kill and process the rattlesnakes and to attract the inquisitive spectators. Venom extraction equipment is at one end of the area, with a chopping block conveniently located next to it for beheading the snakes. Handlers make a point of showing the hypodermic fangs to the audience prior to milking and beheading.

The dangerous process of venom milking begins by using a J-hook to "pin" the rattlesnake's head, or hold it down firmly to a solid surface, so as to be able to grasp and immobilize it. The snake is forced to bare its fangs by applying pressure that opens the upper and lower jaws, and then the upper jaw is pressed against the edge of a glass jar (fig. 17). The milking consists of applying pressure to the venom gland in the snake's upper jaw. Each milked snake expels only a few milliliters of venom into the jar.

Based on venom price lists from two commercial venom laboratories and prices given to us by a venom dealer, freeze-dried *Crotalus atrox* venom is sold for $45 to $125 per gram. One major venom collector/processor/dealer told us that "most of the world's supply of *Crotalus atrox* venom comes from the Texas rattlesnake roundups. It is the most commonly used venom in the world for antivenin production and research" (K. Darnell, pers. comm.). It seems reasonable to assume that if one person takes the time and effort to milk 1.6 liters (1,600 cc) of venom out of western diamond-backed rattlesnakes brought to one roundup, there must be a market for it somewhere. However, we found that details about the business arrangements between venom dealers and antivenin producers are well guarded by both sources.

Skinning and butchering equipment consists of a stainless steel table

Figure 17. Venom extraction at the Sweetwater roundup in 1991 (by Clark E. Adams).

and overhead rack (fig. 18). Processors use the rack to hang up the dead snakes for disembowelment and skinning. After a snake is tied, tail-end first, to the overhead rack, its belly is slit lengthwise. The entrails are emptied onto the table and later pushed into a five-gallon bucket. A processor then strips the skin from the snake. This is one of the most popular shows at the Sweetwater roundup. People line up five and six deep to watch the processing. Many appear to be mesmerized by the milking, decapitation, and processing of the rattlesnakes.

Very little of a butchered snake is wasted. Roundup organizers collect and sell the unprocessed parts, including the decapitated snake heads ($1 per head), skins ($4 per foot),

and rattles ($1 a set) to dealers, taxidermists, vendors, and occasionally the public. A museum curator from Denver, Colorado, purchased fresh skins to be used to construct Native American artifacts for a display. Snake livers and entrails are not usually sold but are collected and used for fish bait by the local sportsmen. Occasionally, Asian-American patrons buy gallbladders for a dollar apiece. Asian folklore claims that snake gallbladders are an excellent human aphrodisiac when consumed raw in an alcoholic beverage.

All the roundups have a fourth area where rattlesnake meat is cooked and sold to the public. This section is usually located away from the rest, sometimes in another building, because of the public danger associated with propane stoves and hot cooking oil. The six-inch-long "chicken-fried" meat strips cost about a dollar per sample. Deep-fat fried rattlesnake meat is very bony. There is not much of it on a strip, and the morsels of meat are greasy. Rattlesnake-meat connoisseurs in West Texas prefer to prepare and serve the meat by first boiling it, then picking the cooked meat from the bones, and finally mixing it with chili. However, urban restaurants such as Arizona's Cowboy Club in Sedona and Rustler's Roost in Phoenix prepare rattlesnake cuisine with more flare and imagination: rattlesnake brochette and rattlesnake-morel tamales, for example.

Elsewhere, rattlesnake cakes are on the menus of the Café Diablo in Torrey, Utah, and at the Seasons Restaurant at Seeley Lake, Montana.

The fifth roundup area is for vendors, exhibitors, advertisers, and concession stands (fig. 19). Some vendors are local groups and businesses, while others travel the roundup circuit attending all or selected roundups to sell their wares. The number of vendors attending a roundup varies from twenty-five to fifty. Fees for floor space depend on location in a roundup facility and the notoriety of the roundup. These fees range from $50 to $75 for a 100-square-foot area at the Taylor roundup to $300 for the same space at Sweetwater. Booths

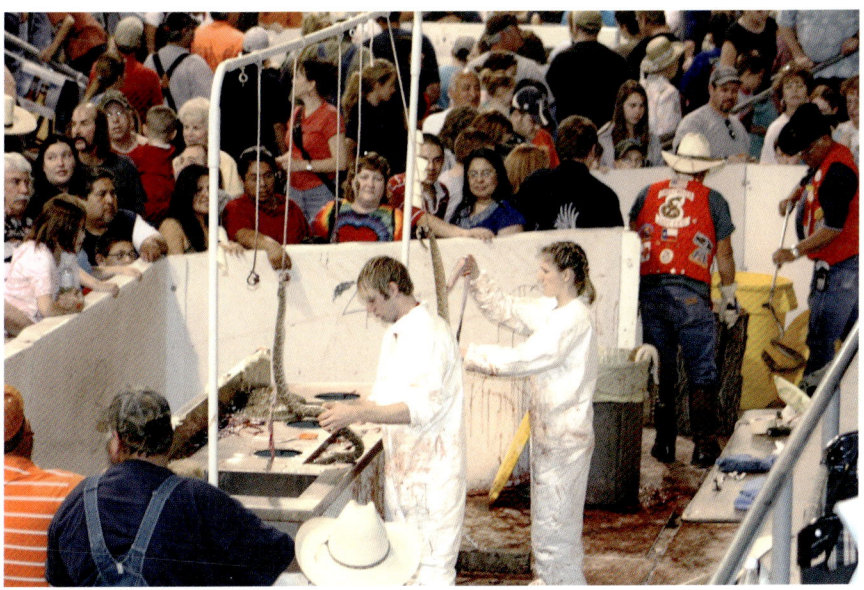

Figure 18. Demonstration of snake decapitation, skinning, and eviscerating at the Sweetwater roundup in 2006, one of the more popular exhibits visited by spectators (by Clark E. Adams).

Figure 19. A typical vendor's table of products made from western diamond-backed rattlesnake parts (by Clark E. Adams).

located near the center of roundup facilities are priced higher than those at the periphery. Vendors' products and prices are discussed later in this chapter.

The sixth area at a roundup features contests as entertainment events. Organizers use some of these events to give their roundup a special identity. For example, the Taylor roundup claims to host the National Rattlesnake Sacking Championship, as shown in figure 20. In this contest, teams of professional snake handlers, and sometimes teams from spectator groups, compete to place ten rattlesnakes in a burlap sack in the least amount of time. One team member holds the sack open while the other collects the snakes from the floor of the exhibit ring and throws them into the opened bag. Obviously, the handler holding the burlap bag is at some risk of being bitten by a rattlesnake that is errantly tossed by his or her partner. Watching the contest gives the phrase "left holding the bag" a whole new meaning.

The Brownwood roundup has featured the "International Championship Rattlesnake Races" (fig. 21). This was the only event of the five roundups we visited in 1991 that involved public participation. Four teams consisting of two members each were selected from the audience. Each person was dressed with leggings made from tire inner tubes to protect against snake strikes. The four-lane raceway was approximately thirty feet long and was enclosed on three sides by a two-foot-high wall. Each team had a racing lane with a box-covered rattlesnake at the starting line. The box served to contain the snake and to keep it calm. When the race started, the snakes were uncovered. Team members blew through long PVC tubes or tapped the tube on the floor behind the snake to encourage it to move down the lane. Teams competed in several heats to determine the overall winning team. The walls along the raceway did not seem to be high enough to protect curious and excited spectators from being bitten eventually by the "racing" rattlesnakes. Although professional snake handlers were nearby, and there were no reported incidents of contestant snake bite until recently, these races are potentially dangerous, as was proven in the 2006 roundup. As mentioned earlier, an experienced, albeit careless, snake handler received a serious bite. The races were not held at the 2007 Brownwood

Figure 20. The National Rattlesnake Sacking Championship at Taylor, Texas (by Taylor Daily Press).

rattlesnake roundup. It remains unknown whether the races will ever be held again.

In addition to these contests, roundups feature other events such as beauty contests, karate demonstrations, square dancing, skits, and ethnic musical entertainment with local residents and invited groups participating. Miss Snake Charmer is a special feature of the Sweetwater roundup, in which young women compete for the crown and prize money. All contestants and the eventual Miss Snake Charmer are required to perform several duties, including skinning and cleaning a dead rattlesnake (fig. 22). There have been forty-four Miss Snake Charmers since 1960. There was no record of a 2003 Miss Snake Charmer in the special roundup edition of the local newspaper, the *Sweetwater Reporter*. Ten young women competed for the

Figure 21. The International Championship Rattle-snake Races in Brownwood (by Clark E. Adams).

2006 crown. Sweetwater also elected a Miss Teen Snake Charmer in 1993 and 1994 but later discontinued this competition.

The Freer roundup provides the most diverse types of entertainment and bills itself as the "Biggest Party" in Texas. Freer organizers enhance the family appeal and marketability of their roundups with more cross-cultural events and groups. These groups include mariachi groups, Native American dancers (fig. 23), and country and western bands in addition to the standard carnival environment of rides, mechanical bulls, and vendors. The color of the costumes, the talent of the young performers, and the quality of professional entertainers who performed there were impressive. These events occurred within a few dozen yards of the snake-holding pits containing some of the largest western diamond-backed rattlesnakes that Texas had to offer. In 1997 the roundup moved from its tent at Rotary Park into a covered rodeo arena at the Cactus Corral grounds. Music headliners for the 2006 roundup were Inzendio,

Figure 22. Miss 1991 Sweetwater Snake Charmer preparing to perform an expected duty of office by skinning and butchering a western diamond-backed rattlesnake. Even in 2006, this was still an expected duty by Miss Snake Charmer (by John K. Thomas).

Figure 23. Native American dance at 1991 Freer roundup (by John K. Thomas).

Conjunto Oro, Los Palominos, Kelly Kenning, and Pauline Reese.

A final area in a roundup facility includes bleachers and a demonstration pit for conducting the public education and/or daredevil programs (fig. 24). At Brownwood and Sweetwater, the 1991 education programs about rattlesnakes were genuinely intended for that purpose and did not involve high risk or sensational tricks, such as kissing a cobra on its head. A change in the snake handlers' club at Brownwood in 2006 shifted the emphasis of the education program to a daredevil show.

At other roundups it was difficult to distinguish between education and daredevil entertainment. Daredevil programs are designed to thrill spectators with risky tricks and events involving rattlesnakes and other snakes (fig. 25). These programs are conducted by members of snake handlers' clubs contracted by roundup organizers. Daredevil performers often include educational information in their shows, both as a public service and as justification for the risky handling of rattlesnakes.

The educational content of these programs was identified by record-

Figure 24. Bleachers and a pit area for snake handler demonstrations is a feature at most roundups (by John K. Thomas).

Figure 25. Typical daredevil show put on by snake handler club members. This individual appeared to be in complete control of the snake (by Clark E. Adams).

ing them at the 1991 roundup sites visited and then reviewing the videotapes. This process revealed that educational programs included seven broad topics: snake identification, correction of myths among the public, rattlesnake behavior, ecological value of rattlesnakes, rattlesnake anatomy, rattlesnake natural history, and public safety. Brownwood, Sweetwater, and Freer conducted the most comprehensive educational programs by addressing most of these topics. All of the programs included the unique features of a western diamonded-backed rattlesnake. Some programs distinguished it from other indigenous, nonvenomous snakes. Some programs attempted to correct various myths about rattlesnakes, including whether the number of rattles indicate the age of a snake, whether a rattlesnake can strike across a distance greater than its body length, if rattlesnakes chase people, and if rattlesnakes lay eggs like chickens. All programs presented information on the commercial and food value of rattlesnakes, but only one covered the rattlesnake's predatory role in nature.

Public safety information was an important part of the educational programs. The frequent interaction between snake handlers and the snakes during a roundup can easily be misinterpreted as less dangerous than it actually is. Consequently, organizers are obligated to instruct the public, particularly children and adolescents, on the safe handling of rattlesnakes and the proper procedures for treating snakebites (see First Aid for Snake Bites sidebar, chapter 1). The most common topics included during safety presentations were recognition of venomous snakes, proper human behavior when encountering a rattlesnake, how to use the syringe venom extractor, and the need for hospital attention. Some safety programs also addressed the myths concerning cutting a snakebite injury, sucking the venom from a wound, and applying tight tourniquets above wounds.

At the 2006 Freer roundup, there was a different approach to public education. Steve Rains, a long-time snake handler and dealer, used an exhibit trailer to provide information on different reptiles, including their habitat requirements and natural history characteristics (fig. 26). He travels a circuit of schools, churches, and communities with his trailer throughout the year. In contrast, there has never been any investment of time or resources by protest groups or members of herpetological organizations to provide public education on rattlesnakes at roundups.

Spectators

Besides the rattlesnakes, spectators are the lifeblood of a community roundup in two ways. They buy the tickets and concession goods, and they draw attention to a roundup with their publicity; even bad pub-

Figure 26. Trailer used to transport a road show on reptiles to schools and communities throughout Texas and other states (by Jessica Alderson).

licity can be good publicity. People attend roundups for a variety of reasons—to satisfy their curiosity about rattlesnakes, for fun and entertainment, to protest the cruel treatment of western diamond-backed rattlesnakes, or to gather information to write articles and books about the event. Roundup spectators consist of local residents, hunters, tourists, vendors, the media, and members of environmental protest groups.

Personal interviews with spectators in 1991 provided important information about their socioeconomic backgrounds, history of attendance, knowledge about rattlesnakes, and attitudes about the impact of roundups on rattlesnake populations. Among the 1,336 people interviewed, or about 2 percent of the estimated attendance at five different roundups in 1991, we found that:

1. Most of the spectators—66 percent—were males.

2. A majority, 63 percent, were white, compared to 36 percent

who were Hispanic. The Freer roundup in South Texas had the largest Hispanic attendance. The Taylor roundup in Central Texas had the largest African American attendance.

3. More than half, 55 percent, were between the ages of 25 and 44, while 11 percent were older than 55.

4. Most were from smaller communities: 33 percent from communities with populations greater than 50,000; 40 percent from communities with populations between 2,500 and 50,000; and 28 percent from small towns with less than 2,500 population, classified as "rural."

5. Spectators traveled an average of 175 miles to attend a roundup, except for the Sweetwater roundup, which had spectators who traveled an average of 275 miles to attend.

6. In occupation, 33 percent of the roundup spectators were retired, a housewife, or a student; 17 percent were employed as professionals or managers; 23 percent were in manufacturing and service-related jobs; and 4 percent farmed.

7. Between 49 percent and 67 percent of the spectators, depending on the roundup location, had attended only one roundup in their lifetime.

Our personal interviews also included a section to determine what spectators knew about rattlesnake behavior, anatomy, and natural history. Interestingly, roundup location and the type of educational program did not seem to produce any obvious differences in spectators' responses to this part of the interview. Most (88 percent) of the spectators knew that rattlesnakes eat rodents. But fewer than one in five people were able to identify other animals included in a rattlesnake's diet, and one in ten people had no idea what rattlesnakes eat.

In response to other questions, (see appendix), many spectators were not well informed about other aspects of rattlesnake mythology, natural history, anatomy, and impact on humans. Fifty percent or more of the spectators, regardless of which roundup they had attended, mistakenly believed that rattlesnakes lay eggs (see fig. 7 for a reason why), that age can be determined by the number of tail rattles; that snakes have eyelids; and that baby rattlers are more dangerous than adults. These misconceptions were rarely addressed during the 1991 public education programs at the five roundups attended. Furthermore, except at Sweetwater, between 67 percent and 88 percent of roundup spectators did not attend the "educational" programs conducted by snake handlers' clubs. It is possible that when asked the question, "Have you attended the educational program at this roundup?" respondents did not equate the snake handlers' shows with an educational program. We suspected that spectators left the roundup with the same levels of knowledge about rattlesnakes as when they arrived, whether or not they attended these shows. A later study of 93 spectators in 2000 by Fitzgerald and Painter resulted in the same conclusion. Finally, we asked spectators at all five roundups to express their opinions on the impact of roundups on rattlesnake populations. Most (73 percent) felt that roundups had minimal or no effect on population size but did control the number of rattlesnakes (84 percent). Nine out of every ten spectators felt that roundups did not threaten the future existence of rattlesnake populations in Texas. Fitzgerald and Painter found nearly identical results in their 2000 study.

Rattlesnake Hunters

The methods of field hunting rattlesnakes in North and South Texas are different. In North and Central Texas, rattlesnakes congregate in dens, whereas the snakes in the warmer climate of South Texas do not. Since 97 percent of the property in Texas is privately owned, all hunting occurs on land owned by the hunter or the hunter's friends and relatives or is accessed by asking permission from a landowner. Landowners often invite hunters onto their property, especially when roundups are conducted in nearby communities.

Some rattlesnake hunters have a generational claim on rattlesnake hunting in Texas. Clifford Etheredge, a cotton farmer and self-proclaimed snake wrangler, provides an interesting account in *Texas Rattlesnake Hunting* of his family's snake hunting experiences over the past one hundred years. He reminisces about the early days of the Sweetwater roundup and spins a few humorous yarns about snake bitten folks. Etheredge was born and raised in West Texas in what he calls the heart of rattlesnake country. Rattlesnake hunting was considered a necessity in order to protect humans and livestock on his family's ranch. It became lucrative for three reasons: (1) the federal government subsidized the collection of snake venom for the production of antivenin in the 1950s; (2) some roundups began at that time to create markets for rattlesnake products, and (3) prize money was made available for the hunters participating in the roundups. These monetary incentives caused a rapid increase in the number of hunters.

Prior to our 1991 study of rattlesnake roundups, there had never been an in-depth examination of the attitudes, activities, and knowledge of rattlesnake hunters. Our team conducted personal interviews with

212 rattlesnake hunters who brought snakes to five different roundups in 1991 (see Hunters in appendix for sampling procedures). The hunters' answers revealed that:

1. They had participated in roundups for an average of nine years.

2. Most were white (76 percent) or Hispanic (23 percent) males.

3. They were between the ages of 18 and 54.

4. Many (45 percent) lived on farms or ranches or in rural areas, or in small towns (27 percent); fewer were from large cities (28 percent).

Their reasons for hunting such a dangerous animal were to gain experience, earn additional income, and reduce the number rattlesnakes in their area.

We also asked hunters about various aspects of rattlesnakes' behavior, anatomy, and natural history (see appendix). It seemed reasonable to expect that hunters would be well informed about the species as a result of hunting it. When asked what rattlesnakes eat, nearly all (94 percent) of the hunters knew that rattlesnakes ate rodents. Only one hunter in ten was able to identify other animals, except for birds, that are included in a rattlesnake's diet. Two percent had no idea what rattlesnakes eat. Nearly 75 percent incorrectly believed that a baby rattlesnake's bite was more dangerous than an adult's bite. Overall, hunters scored marginally higher (9 percent) than did spectators on questions about rattlesnake behavior, anatomy, and natural history.

When hunters were asked to express their opinions on the impact of roundups on rattlesnake populations, 60 percent said minimal or no effect, 31 percent said hunting rattlesnakes for roundups helps control population size, and only one in ten hunters felt that hunting threatens the snakes' future existence.

Rattlesnake Hunts

Most hunters in North Texas hunted in teams of two or more members because of the rugged terrain and often precarious location of dens. As shown in figure 27, many of these dens are located in the sides of hills or embankments that require hunters to climb or use ladders to access the dens. At a rattlesnake den site, a member of the team first looks into the den using a hand-held mirror in order to detect the presence of rattlesnakes at the entrance or inside. Next, a long copper tube is inserted as far back into the den as possible. The tube is connected to a spray can containing a highly volatile gasoline. Various quantities of gasoline fumes are sprayed into the den to flush out the resident rattlesnakes. One member uses a J-stick to restrain each snake that exits the den while the other member holds the capture bag. When most of the emerging snakes have been captured or have escaped capture, the hunting team members gather up all their equipment and move on to another den or quit for the day.

The J-stick used to restrain rattlesnakes is similar to the tool used to pick up roadside litter. The pressure of the pincher end of the snake stick is controlled by the hunter's hand and the handle. The more hand pressure is applied to the handle, the tighter the pincher clamps down on the snake's body. Needless to say, the adrenaline flow in an inexperienced hunter attempting to restrain a buzzing, wriggling rattlesnake may cause such a hunter to squeeze down too hard on the handle, thus damaging the delicate skeleton and internal organs of the snake. However, experienced snake handlers, as shown in figure 28, know how to grasp the rattlesnake properly with a J-stick in manner that protects both from danger during the encounter.

Using gasoline fumes to flush rattlesnakes out of their dens has raised the greatest level of controversy about rattlesnake roundups. Ecologists and herpetologists have grave concerns that the gasoline method used to drive snakes out of their dens will produce serious side effects on the rattlesnakes and other den occupants. Jonathan Campbell, a university professor of herpetology, found that if you exposed western diamond-backed rattle snakes to gasoline fumes for 30 to 60 minutes in a laboratory test (gas) chamber, the surviving snakes needed long recovery times (one to seven hours) to perform normal tongue-flicking, the righting response, and tail rattling behavior. One snake died after 60 minutes of exposure. Campbell also found that exposing other species of snakes and lizards to 30 to 60 minutes of toxic gasoline fumes produced similar detrimental effects. It should be pointed out that his study did not replicate field conditions associated with spraying gas into an open den. The animals in the den were not confined to exposure for 30 to 60 minutes in a closed vessel. They had the opportunity to escape, and the fumes probably dissipated in a shorter

Figure 27. Dens are gassed to drive out western diamond-backed rattlesnakes. Thin copper tubing is being inserted into the den (by John K. Thomas).

Figure 28. Holding a rather large rattlesnake with a J-stick (by Richard Delifka).

period of time than was the case in the laboratory test chamber.

There has been only one field study (Speake and Mount 1973), conducted outside Texas, where it was shown that gassing eastern diamond-backed rattlesnakes residing in gopher tortoise burrows had some "possible ecological effects" on the snakes and other species. Gopher tortoise burrows can be six to eight feet deep and thirty feet long. In order to accomplish this study, seventy-four gopher tortoise burrows were "excavated," meaning permanently destroyed; recall that the American Society of Ichthyologists and Herpetologists commented on "degradation of habitat" (see position statement sidebar, chapter 2). The study revealed that of the fourteen eastern diamond-backed rattlesnakes subjected to the treatment (gassed), some lived, some died, and some had temporary behavioral problems.

In both of the studies mentioned, the researchers killed some animals with gasoline fumes to demonstrate that gasoline fumes kill animals. Conclusions were extrapolations of these results: if they could kill rattlesnakes and other animals with gas fumes, then hunters using gas to drive rattlesnakes from dens would also kill snakes and other animals. The public expects the scientific community to provide credible tests of cause and effect using rigorous scientific procedures. Common sense notwithstanding, one indication of scientific credibility occurs when studies appear in the high-visibility journals of the scientist's discipline; these studies did not.

In order to provide a credible scientific test of the effects of gas fumes in rattlesnake dens, the scientific community has always faced somewhat of a conundrum. First, a credible scientific investigation must be conducted under strict field conditions, which would be difficult. Second, it is doubtful that the study would be approved by the university animal use and care committees established to provide oversight on investigations involving animals. One can easily imagine the concern for academic integrity and quality of science by committee members considering a research proposal that exposes animals to gasoline fumes until they suffer behavioral impairments or die. Third, it is doubtful that competent herpetologists would want to be associated with this kind of study. The reaction of one highly published wildlife biologist to the two studies mentioned was that "some studies do not have to be conducted when the expected outcomes are self-evident" (M. Peterson, pers. comm.).

Hunters have long known that exposure to gasoline fumes affects other species cohabitating in the dens. Clifford Etheredge's account confirmed the flight of mammals and reptiles from gassed dens during his rattlesnake hunts. He did not provide any additional comment concerning to what degree other species (vertebrates and invertebrates) were harmed or killed during the den-gassing process. Hunters told us that they have gassed and harvested rattlesnakes from the same dens for years. One hunter reported that he had hunted at the same den for twenty-nine years. One professional snake handler believed the gasoline "sterilized" the den of rattlesnake parasites that cause more harm to the snakes than do gasoline vapors.

Finally, hunters who gas a den may get more than they bargained for. One story illustrated the fire danger associated with gassing dens. A careless hunter flipped a burning cigarette, which landed near the entrance of the den that had earlier been gassed. The resulting explosion removed a significant portion of the hillside, and his truck was badly damaged. Fire danger aside, a bunch of angry snakes in a gassed den can spill out all at once, overwhelming a single hunter with a J-stick and burlap sack.

Hunting in South Texas is less controversial in terms of the method used. Hunters used to "patrol" along public roads looking for snakes that were either basking in the sun or trying to cross the road. As a result of the nongame permit regulations, road hunting now brings a stiff fine in order to discourage it. South Texas hunts occur during more months of the year because of the warm climate, which means snakes spend less time in hibernation and more days in the open hunting prey. Thus finding rattlesnakes six feet long is not uncommon in South Texas.

Hunting for rattlesnakes generally intensifies a few weeks before a roundup. Hunters bring captured rattlesnakes to roundups to compete for prizes for having the largest, longest, heaviest, most, and smallest snakes. They transport their captured snakes from the field in burlap bags, trash cans, or specially constructed boxes measuring two feet wide, four feet long, and six inches high. These boxes have wide lids, permitting easy storage and retrieval of snakes (fig. 29). The low height of the box prevents the snakes from compacting and crushing other snakes lying on the bottom. Some

Figure 29. A common method of transporting captured western diamond-backed rattlesnakes to and from roundups (by Clark E. Adams).

professional hunters and most of the dealers use these boxes to hold their snakes in captivity.

How long hunters hold rattlesnakes in captivity before delivering them to a roundup site is unknown. It is reasonable to suspect that most hunters do not keep rattlesnakes in captivity very long, given reports that they hunt for only a few weeks prior to a roundup. Captured snakes lose weight, suffer, or die from the effects of long-term crowding in their container. One hunter kept hundreds of rattlesnakes in a secured chicken house throughout the winter until they could be sold at the Sweetwater roundup in March. He tried to feed them mice, but the rattlesnakes were not receptive to the offer. Since winter is the normal hibernation time for rattlesnakes in North Texas, perhaps they were more in the mood for sleeping rather than eating.

Whatever storage method is used, hunters are aware that dealers will buy only healthy, live snakes at roundups. Injured and dead rattlesnakes rapidly deteriorate, affecting the quality of their skin, meat, and other commercially valuable parts.

At most roundups, hunters deliver snakes to an area located away from the spectators. However, at one roundup, hunters had to walk through the crowd of spectators to deliver their snakes. Neither hunters nor people in the crowd seemed too concerned about the potential danger this posed. They delivered their captured snakes to the holding pits where roundup personnel weigh the snakes. The snakes were then emptied from their containers onto the floor of the holding pit. The inability of the rattlesnakes to escape their situation causes them to curl up, each forming a tight bundle and hiding its head among the coils of its body. This curling is the same behavior the snakes exhibit when confronted with a potential predator, such as a hawk soaring overhead. Snake handlers call this behavioral response the "pancake" or "cow patty" rattlesnake, as illustrated in figure 30.

After many conversations with rattlesnake roundup hunters and organizers, we found a glaring contradiction in their justification for hunting rattlesnakes for roundups. On the one hand, they claim that by removing rattlesnakes they are protecting family, pets, and livestock from rattlesnake bites. On the other hand, they say that hunting rattlesnakes for roundups does not necessarily hurt the rattlesnake population. To accomplish the first task of protecting family, pets, and livestock, they would have to deplete the rattlesnakes to some level below what was originally there. If hunting rattlesnakes is not decreasing the size of the rattlesnake population, as they claim, then how is hunting protecting family, pets, and livestock?

Vendors, Products, and Exhibitors

Roundups have an interesting mixture of vendors and exhibitors, including military recruiters. Rattlesnake dealers are also vendors at all the roundups, selling a variety of rattlesnake-related products (fig. 19). One vendor, George Wills, expressed in 1994 his regard for consumers: "We thank God for the Yankees who come down here to help us out and buy our rattlesnake products."

Figure 30. The so-called pancake or cow patty snake that can be held by one of the spectators (by Clark E. Adams).

We have already noted the marvelous anatomical adaptations that the western diamond-backed rattlesnake has for survival in its natural habitats. This section gives an accounting of the economic value of these physical adaptations or body parts in the marketplace. The diversity of products is extensive, but a general summary of what can be purchased at nearly any rattlesnake roundup is given in table 3. There was also the "Himalayan rattlesnake" spoof. Himalayan rattlesnakes are taxidermy mounts of western diamond-backed rattlesnakes coated in white rabbit fur. We did not query spectators to determine who believed that such a creature actually exists.

If customers did not prefer any of the rattlesnake products, they could also purchase a mounted alligator head ($200), an alligator back-scratcher ($8), or a mounted armadillo dressed up like a cowboy ($100).

The most bizarre product was called a "raccoon ear bone," which sold for $4. An ear bone is actually the penis bone (called the baculum) of a male raccoon. One has to wonder why someone would buy this oddity and what they would do with it.

There is also an assortment of other vendors who are outside the rattlesnake trade. They sell more conventional products such as AKC registered dogs, food and beverages, arts and crafts, camping gear, cellular phones and services, clothing, knives, and an assortment of wildlife products. The alcoholic beverage industry likewise capitalizes on the roundup market with promotion and sale of popular brands of beer (Budweiser, Coors, and Miller Lite are most prominent). Although calling the military

TABLE 3. SUMMARY OF THE PRODUCTS PRODUCED FROM RATTLESNAKE
PARTS THAT CAN BE PURCHASED AT RATTLESNAKE ROUNDUPS

Apparel and accessory products made with western diamond-backed rattlesnake skins	Price
Bikini	$200
Hat band	$40
Clutch purse	$30
Wallet	$18 to $20
Check book cover	$20
Money clip	$12
Belt	$20
Belt buckle	$15
Rattle necklace	$13
Rattle earrings	$10/pair
Head bands	$6
Coin purses (large and small)	$10 and $5
Snake head key chain	$5

Mounted snakes and related artifacts	Price
Coiled rattlesnake with extended fangs	$40
Snake patties (see figure 30)	$12
Large pickled rattlesnake head	$15
Skin with rattle	$35
Skin without rattle	$8
Skin with vinyl backing	$45
Golf balls covered with rattlesnake skin	$7/pair
Turtle shells with rattlesnake heads	$25
Baby rattlesnakes seemingly hatching from duck or chicken eggs (see figure 7)	$10

branches vendors may be stretching the point, army, marine, and national guard recruiting booths were present at the 2006 roundups. We can only speculate that the military services may have considered some roundup participants to be the "right stuff" as recruits. In 2006 the armed services were motivated by the need to recruit more young men and women to serve in the Iraq war.

Snake Handlers

One of the first snake handlers and showmen in Texas (circa 1907) was Willie Liberman, who adopted the trade name of W. A. Snake King. He established one of the first snake farms, called Snakeville, in Brownsville, Texas. In 1914 he organized the first rattlesnake-catching contest during the Sixth Annual Mid-Winter Fair in Brownsville. He rode a donkey four hundred miles to Austin to invite Governor O. B. Colquitt to the fair and contest personally. The governor accepted the invitation. Mr. and Mrs. W. A. Snake King were the uncontested winners of the first rattlesnake-catching contest. Today, this contest is called the sacking contest and it is usually an integral part of the Taylor rattlesnake roundup.

Snake handler clubs were first organized during the 1950s. The Venomaires from East Texas came first. This group disbanded later and the Heart of Texas Snake Handlers

Club was organized in 1975. Seven snake handlers' clubs were operating in Texas during the 1990s: the Heart of Texas Snake Handlers, Texas State Diamondback Hunters, Snakes Unlimited, Sandhills Rattlesnake Club, South Texas Snake Handlers, Fangs and Rattlers, and Rattlers Plus Snake Handlers. The seven clubs had a total of fifty-six members, primarily white (55 percent) males (48 percent). The only nonwhite member was a Hispanic male. Today there are four rattlesnake handler clubs left in Texas: Heart of Texas Snake Handlers, South Texas Snake Handlers, Fangs and Rattlers, and Texas Diamond Back Handlers.

As discussed previously, their roundup programs focus on a blend of safety, natural history information on rattlesnakes, and/or daredevil entertainment. In addition to roundups, the snake handler clubs perform for gun shows, charities, private business events, grade schools, 4-H clubs, Boy Scouts, and Cub Scouts. The number of shows at roundups varies from one to eight daily, each lasting from thirty-five to forty-five minutes. A club typically performs for less than twenty events annually. Three clubs limit their events to Texas, while the others perform throughout the South and Southeast.

Roundup organizers contract with a snake handler club to conduct daredevil shows. A ticket sold for one dollar in 1991 and five dollars in 2006. Some of the acts include holding coiled rattlesnakes in their bare hands (fig. 30), getting into a sleeping bag with dozens of live snakes (fig. 31), provoking rattlesnakes to strike balloons and other objects, holding the tails of six rattlesnakes in each hand (fig. 32), and seemingly mesmerizing a cobra and kissing its head (fig. 33). All of the handlers have several years of professional experience, and some have scars confirming that they know the agony of snake bites. Unlike in 1991, the 2006 Freer

roundup was based almost entirely on non-snake-related entertainment, including country and western music concerts, dance performances, theatrical skits, and carnival rides. Many of the scheduled daredevil shows at Freer were cancelled due to lack of public attendance.

Figure 31. A common trick used in daredevil shows consists of placing a dozen or more snakes into a sleeping bag occupied by a person. The trick is to use lots of snakes that will be more attracted to one another than to the person in the bag (by Clark E. Adams).

Figure 32. Holding six western diamond-backed rattlesnakes (a six-pack) in each hand. Needless to say, the snakes are not very happy and tend to strike at one another in this position (by Clark E. Adams).

Figure 33. Kissing the head of a cobra is a trick that takes advantage of the cobra's downward strike, as opposed to the forward striking behavior of the western diamond-backed rattlesnake (by Kenneth Drum).

Snake handlers have different reasons for participating in public entertainment involving rattlesnakes. One fellow summed it up by saying, "We are all nuts!" Another said it was a hobby that gave him relief from his otherwise stressful job. Another reason given was to educate the public about rattlesnakes.

All handlers acknowledge the risks involved in handling rattlesnakes but consider the dangers to be part of the challenge of mastering this venomous species. However, on occasion the rattlesnakes have the upper hand (no pun intended)—snake handlers, both men and women, display their

scars on limbs and digits from previous snake bites (fig. 34). A few reported having large ($40,000) outstanding hospital bills for previous snake bite treatment. Needless to say, snake handlers have difficulty getting hospital insurance because of snake bite–related injuries.

Bill Ransberger of Sweetwater is considered one of the deans of professional snake handlers and educators at rattlesnake roundups (fig. 24). As noted, his credentials include the claim that he has been bitten forty-two times by rattlesnakes and lived to recount his long career of experiences. He has probably trained more

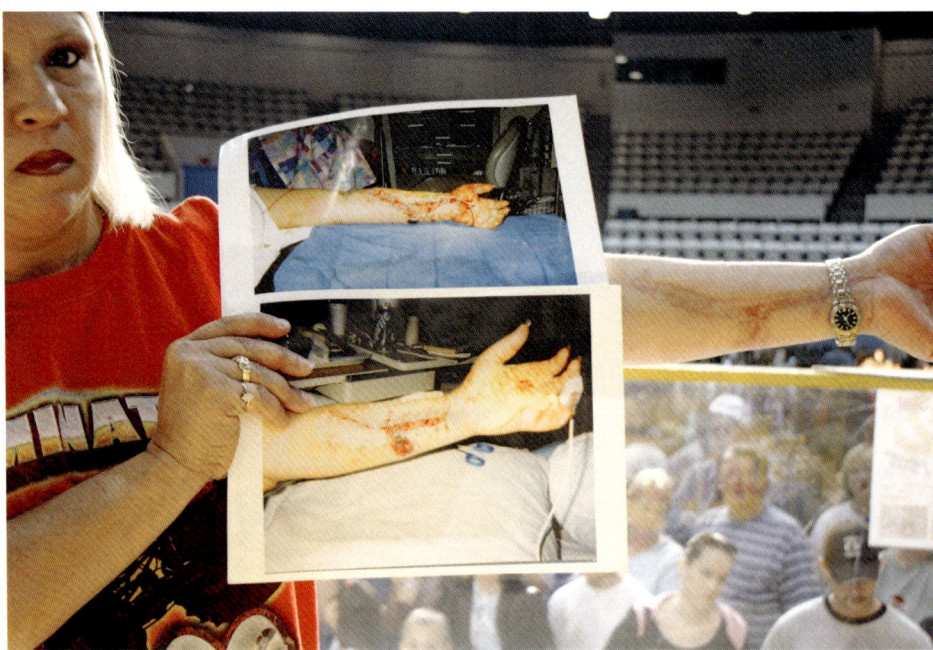

Figure 34. Sometimes a rattlesnake bite can lead to serious bodily damage and a long recuperative time (by Clark E. Adams).

people in proper handling of venomous snakes than anyone. Ransberger began doing roundups in 1958 and was instrumental in organizing the Sweetwater, San Angelo, Big Spring, and Andrews roundups. In a March 2000 interview with a newspaper reporter, he was asked to provide some safety tips about what to do when confronted by a rattlesnake. Ransberger advised that the best things a person can do to avoid being bitten are to be sure not to make any quick moves, determine where the snake is located, and then slowly back away from it. He retired from organizing roundups in 1997, but his legacy continues in Sweetwater.

The story of snake handlers at rattlesnake roundups would not be complete without some mention of Jackie Bibby (fig. 35; http://www.texsnakeman.com). He is proud of his place in four *Guinness World Records* editions (2005, 2006, 2007, and 2008) for occupying a sleeping bag with 112 rattlers and a bathtub with 84 rattlesnakes. He has another world record for holding eleven snakes, by their tails, in his mouth. Recently he achieved a further world record by being the first person to enter headfirst into a sleeping bag already occupied by twenty rattlesnakes. Bibby began his career as a snake handler in 1969 when he was eighteen years old. He has received eight rattlesnake bites since then. One resulted in the

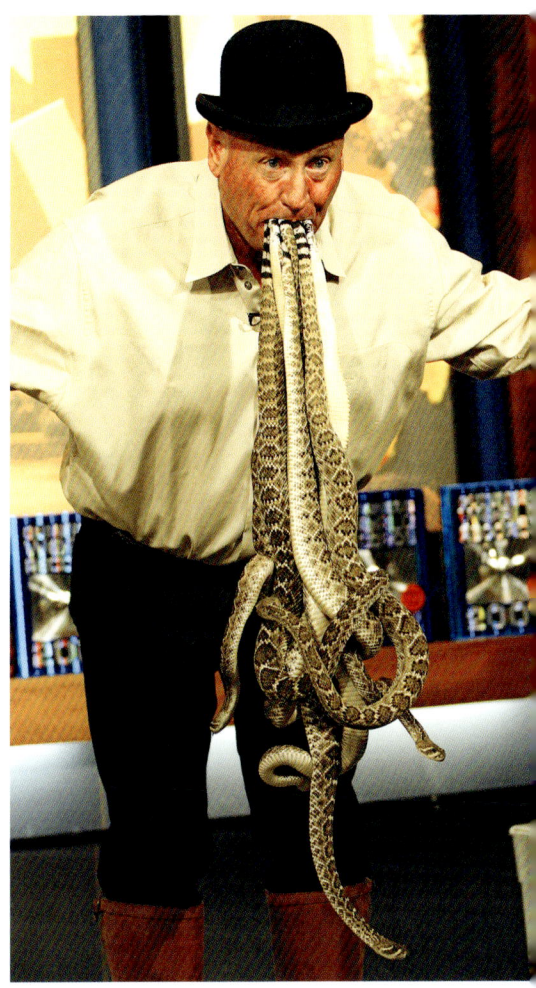

Figure 35. Jackie Bibby demonstrating one of the rarer stunts conducted by snake handlers (used with permission from Guinness World Records 2006).

loss of function in one thumb. The primary reason he has been a snake handler for so long is that he loves the adrenaline rush and the competition (Jackie often says handling snakes in front of thousands of people is the most fun he has ever had with his pants on). Beyond rattlesnake round-ups, he makes his living as director of an outreach center, a recovery-oriented boarding house for people suffering from an addiction to alcohol or drugs. Bibby is himself a recovering addict.

Chapter 4

Impact of Hunting on Rattlesnake Populations

While the lack of scientific data to address the effects of hunting and harvesting on rattlesnake populations has persisted over the past half century, opponents of rattlesnake roundups speculate that the extensive harvest of western diamond-backed rattlesnakes for roundups and commercial skin dealers is depleting rattlesnake populations. One member of the Greater San Antonio Herpetological Society estimated in the 1980s that half a million rattlesnakes were harvested annually in Texas. The accuracy of these claims, made without reliable supporting data, has never been verified.

The harvest of rattlesnakes for a roundup may be influenced by three groups of interrelated factors:

- *Ecological factors*—the weather (e.g., average minimum and maximum temperature and precipitation during February), the number and status of snakes in the natural population.

- *Hunting pressures*—previous year's rattlesnake harvest, time of the year when the roundup was conducted, number of hunters (competition) and days annually spent in the field, and present hunter success.

- *Market factors*—commercial dealers' inventory on-hand, price paid per pound, public demand for rattlesnake products, community support of roundups, and roundup organization.

The degree to which any of these factors affect harvest rates remains unknown because of the complex interrelationships that exist among the factors. It is problematic for field biologists to conduct research on such complex issues because of the extreme difficulty in isolating and measuring how these and other variables, such as collection regulations, impact harvest rates.

Some Long-term Harvest Data

The Sweetwater roundup has reported harvest data (pounds of rattlesnakes) in its local newspaper each year since 1959. Dealers do not usually count the actual number of snakes brought to roundups because they pay hunters by the pound. Therefore, we felt that another way of understanding the harvest or "take" of rattlesnakes for roundups was to find a reliable method of converting the historical records on the total pounds (biomass) of rattlesnakes brought to roundups into estimated numbers of snakes. To aid in this endeavor we found that for the last thirteen years (1994–2006) Sweetwater has been obtaining sex, weight, and length data on randomly drawn samples (a total of 6,060 snakes) of rattlesnakes from those brought to their roundup by hunters. For this period, these data revealed that 75 percent of the snakes were males and 25 percent were females. Their average weights were 2.10 and 1.44 pounds, respectively. Using this information we estimated number of western diamond-backed rattlesnakes brought to the Sweetwater roundup over the thirteen-year period. We also extrapolated backward in time all the way to 1959 when the Sweetwater organizers began recording the rattlesnake biomass brought to their roundup. We did this because statistical analysis of the thirteen years of data showed that the average weight of females and males was approximately the same each year.

From this point, the method of converting biomass into numbers was quite simple. First, the total biomass needed to be divided into two samples (25 percent for females, and 75 percent for males). Next, dividing each sample by the average weight of females and males, respectively, gave the estimated number of both sexes. Adding these numbers gave the total population of western diamond-backed rattlesnakes brought to the Sweetwater roundup during a particular year. This procedure was used to estimate the number for each year and the total from 1959 to 2006. Over the last forty-eight years (1959 to 2006), hunters have brought 140,229 western diamond-backed rattlesnakes to the Sweetwater roundup, as calculated from the total biomass of 264,211 pounds. The average number of snakes over this period was 2,921 each year. The largest harvest for the Sweetwater roundup was 9,546 snakes in 1982. The smallest harvest was 721 snakes in 1987. As illustrated in figure 36, there were extreme variations in the number of rattlesnakes brought to the roundup each year, for reasons given earlier, but there was no significant increase or decrease over the forty-eight-year period. In other words, the high years of harvest were offset by the low years of harvest.

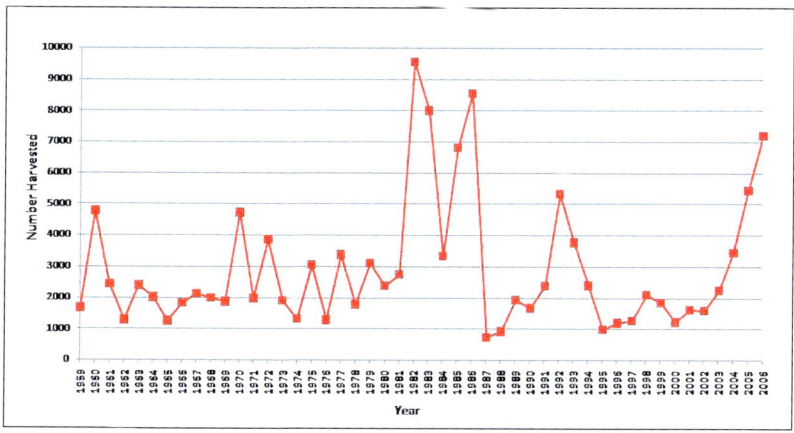

Figure 36. Graph of the number of western diamond-backed rattlesnakes brought to Sweetwater roundup from 1959 to 2006 (by Linda Causey and authors).

What is important to realize is that the harvest of rattlesnakes for the Sweetwater roundup has been a sustainable process for the past forty-eight years. What does this mean in terms of impact on the populations of rattlesnakes hunted for this roundup? We believe that this forty-eight-year harvest has been supported by an annual, sustainable replacement rate in those rattlesnake populations from which the hunters are harvesting rattlesnakes for the Sweetwater roundup. In other words, the hunting of rattlesnakes for this roundup has had a negligible impact on rattlesnake populations. We should add that although Sweetwater is the largest roundup, it represents only part of the roundup and non-roundup harvest of western diamond-backed rattlesnakes in Texas. The Sweetwater numbers, therefore, are a conser-

vative indicator of a larger activity of which we know little. Furthermore, roundup supporters can hardly claim that the removal of nearly ten thousand rattlesnakes in one year (1982, see fig. 36) from the area surrounding Sweetwater did not have any impact on rattlesnake populations.

Geographic Range of Rattlesnake Hunters

We used a second and indirect method to determine the possible impact of rattlesnake hunting for roundups on the Texas rattlesnake population. This was to ask hunters to identify the counties where they hunted. Most restricted their rattlesnake hunting to within a hundred-mile radius of the roundup community where they lived. Figure 37 shows the 83 counties within the 111-county rattlesnake range that hunters reported to have

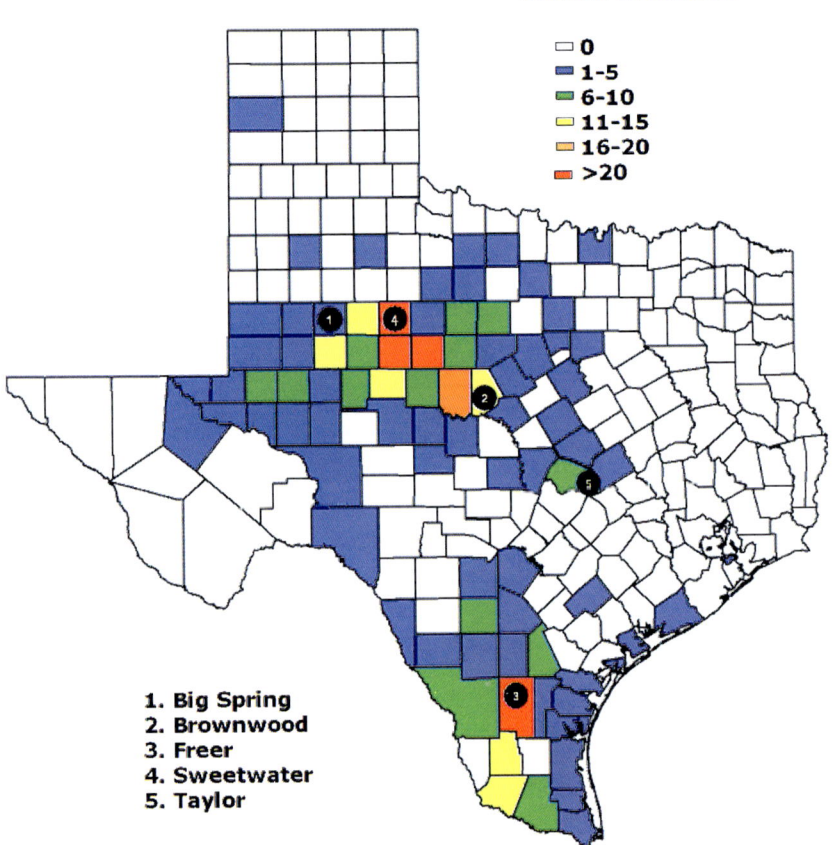

Figure 37. Map of the counties where western diamond-backed rattlesnakes were hunted and the number of hunters per county for the 1991 roundups (by Clark E. Adams).

hunted at least once in 1991. Only four of these counties had twenty or more hunters; seven had between ten and twenty hunters. From 1994 to 2001, the Sweetwater Jaycees found that, on average, hunters used only thirteen counties to hunt western diamond-backed rattlesnakes. Over-

all, much of the hunting effort was in the county where the roundup was located or adjacent counties.

In North Texas, road access to den areas was important because of the heavy equipment needed to spray gasoline into rattlesnake dens. This requirement excluded any dens that were not within easy reach of roads. The resulting limited distribution of hunters and the restrictions on their access to dens indicated that millions of acres of rattlesnake habitat were untouched by hunters participating in rattlesnake roundups. Some ecologists believe that during the hunting season the rattlesnake population is highly clumped, such that even when hunting a small area, hunters could nevertheless take out a large proportion of the population, but there are no data to support this belief.

The geographic range of rattlesnake hunters could be even more constrained today than it was in 1991 for economic reasons: the price of gas increased from 85 cents a gallon in 1991 to about $3 a gallon in 2006, and at the time of writing western diamond-backed rattlesnakes were bringing only about $2.00 per pound, compared to $5.50 per pound in 1991. J. E. Morales, a dealer at the Freer roundup, summed up rattlesnake hunting today in the following way: "There have been a lot of changes over the years. In the beginning of the roundups, every family would get in their trucks and drive the roads at night picking up snakes. Ranchers opened up their gates and gave out maps of their land, allowing people to hunt for snakes on their property. Oilfield workers would collect snakes on their job sites. No one would ever say anything. Today, you cannot get on the road and hunt for snakes from

a vehicle because it is against the law. Many land owners do not want the liability of people hunting snakes on their property." According to Morales, if you get caught picking up a snake on a county road by a game warden, you can be fined $100 to $200. To hunt and sell snakes, you must have a hunting permit and a nongame collector's permit. Many people do not want to go to all that trouble just to hunt snakes, he explained. Snake hunting also costs money. Morales said he could "drive 90 miles on a snake hunting trip. With the current gas prices so high, that can get very expensive."

Any determination of hunting impacts on rattlesnake populations is difficult without a statewide assessment of the size of the western diamond-backed rattlesnake population at a given point in time. This task would require correlating harvest data with rattlesnake census data taken from the field and with experimental data to measure rates of natural attrition and replacement. These rates could vary in different areas of the western diamond-backed rattlesnake's extensive range throughout Texas, compounding the difficulty of the effort. Biologists have discussed the need for such a study, but the cost and the complex logistics of study design prevent the work from actually taking place. No one has ever been able to provide a baseline record of the actual numbers of rattlesnakes in any habitat scenario with which subsequent studies could compare data. In fact, we could not find any literature that provided even small regional population estimates.

Hunting for Commercial Trade

As mentioned earlier, dealers may purchase rattlesnakes before and during the roundups in various communities. Dealers have reported, however, that only 15 percent of their rattlesnake inventories were purchased at roundups. To extrapolate from this percentage in an effort to determine the overall annual rattlesnake harvest by commercial dealers hardly offers much potential for accuracy.

Some dealers travel routes throughout the state and never buy their snakes at the roundups. This often involves their posting notices at feed stores, gas stations, and convenience stores on particular small-town routes. These notices state when the dealers will be in the area, what species they are interested in buying, and how much they will pay. Understandably, dealers are reluctant to provide exact harvest information from all sources.

As a result, any estimates of rattlesnake harvest based solely on roundup data or information from dealers would be conservative at best and downright incorrect in most cases. The real debate should be over the

impact that unregulated commercial exploitation has on western diamond-backed rattlesnake populations.

Field Study of the Impact of Hunting on Rattlesnake Populations

The simple fact that western diamond-backed rattlesnakes play an important part in nature's food web merits consideration of studying and trying to understand human impacts on this species. A field study designed to determine the impact of hunting on rattlesnake populations should address several issues. The first question arising involves the size of the area to be included. The choices are the whole state, regions, the county, the ranch, or the den. The complexity of such studies can increase nearly exponentially with geographic scope; a state-level assessment would be a far cry from den-level analysis and is unlikely. However, long-term studies of population changes at the den level of analysis could be conducted. Hunters said they hunted the same dens year after year and always caught snakes. Yet this kind of study has limitations. Dens come in various locations, shapes, and sizes, which could influence the number of rattlesnakes that live there. Also, counting the number of snakes at a den where they congregate does not indicate what is happening to the population at large. It would be like counting people at a shopping mall and extrapolating to the whole Texas human population.

The next consideration involves methods to be used in order to determine the actual numbers (population size) of rattlesnakes in the study area. Wildlife managers have always struggled over the best methods of determining population size and usually end up relying on some kind of "estimator." Biologists rarely know the actual number of animals within an area, whether they are deer, quail, turkeys, or rattlesnakes. People can readily gauge when wildlife populations are "overabundant," when a species appears everywhere in an area, or "missing," when a species has left an area. Any number between overabundance and missing is an estimate, meaning that the population size of a species in an area can be expressed given a certain level of statistical confidence and an acceptable margin of error.

A field study to determine the size of a rattlesnake population in a den requires that as many snakes as possible be caught, marked, released, and recaptured. However, a researcher would have difficulty capturing all live snakes in every den, making it difficult to conduct a complete census of the population. Some snakes would be caught easily and repeatedly. Others might never be caught. Some would be caught once and never again. Some snakes caught might be temporary visitors not really belonging to the den population. Some might move away and take up residence in a neighboring den. Moreover, lack of accessibility makes many dens and their occupants unknown. All these scenarios make it difficult to determine the actual size of the rattlesnake population even at the den level of analysis. The following discussion describes one method used to estimate population size in spite of the inherent problems of interpreting the meaning of the results.

Population estimates are derived by first using marking techniques to identify each captured snake individually (temporarily or permanently). In the past, marking procedures involved the use of radio transmitters implanted under the snake's skin. Each snake was tracked with a hand-held radio receiver keyed to a unique signal coming from the transmitter. Another marking procedure involves passive integrated transponders (PIT tags) that are injected with a syringe into the subcutaneous cavity of the snake. We could not find any information indicating whether this procedure hurts the snake. Each tag consists of an embedded microchip that transmits a unique identification code. Individual snakes are identified with a hand-held scanner that displays the code. The snake is released back into the den population after capture and subsequent recaptures.

Second and subsequent rounds of captures produce both marked and unmarked snakes, the unmarked ones becoming marked after their first capture. After several rounds of captures and recaptures, a final sample of marked and unmarked snakes is drawn. The population size is estimated by calculating the ratio of marked to unmarked snakes using procedures suggested by Lancia and associates in *Techniques for Wildlife Investigations and Management.* This guide was developed by members of the Wildlife Society and others who are authorities on the various techniques needed to study wildlife populations.

In order to determine the effect of hunting on rattlesnake populations, two types of dens would be needed: dens that are not hunted are called "control" dens, and dens that are hunted regularly are called "experimental" dens. Scientific experiments always use control and experimental groups in order to determine if some type of treatment (in this case hunting) results in measurable population differences. A scientist would look for measurable differences in the number of rattlesnakes that occupy the control and experimental dens over some time period. How many dens would have to be in each group? In experimental design, the more the better; thirty or more would be preferred, thirty being the smallest sample size that might produce statistically significant results. Finding thirty dens might be difficult, given terrain and access conditions. Furthermore, such a study would require several years of repeated measures of the number of rattlesnakes occupying hunted and unhunted dens. This requirement would mean that some dens would be hunted annually, while unhunted dens would need to be protected over the same period.

By now, the complexity of problems associated with a field study to determine the impact of hunting on rattlesnake populations in dens should be becoming apparent. Many aspects of rattlesnake natural history also need to be factored into such a study. For example, biotic potential and environmental resistance issues, mentioned earlier, have to be incorporated into the mix of things being investigated. These might include investigations of how many offspring rattlesnakes can produce each year, what environmental conditions affect their reproductive rate, and what animals prey on rattlesnakes in different geographical and ecological areas of the state. Rattlesnake hunters claim, for example, that there are noticeably fewer rattlesnakes in areas with abundant white-tailed deer and turkey populations. It remains unknown whether this observation is accurate and, if so, what would the reason be. Perhaps deer and turkey hunters in these areas shoot rattlesnakes on sight.

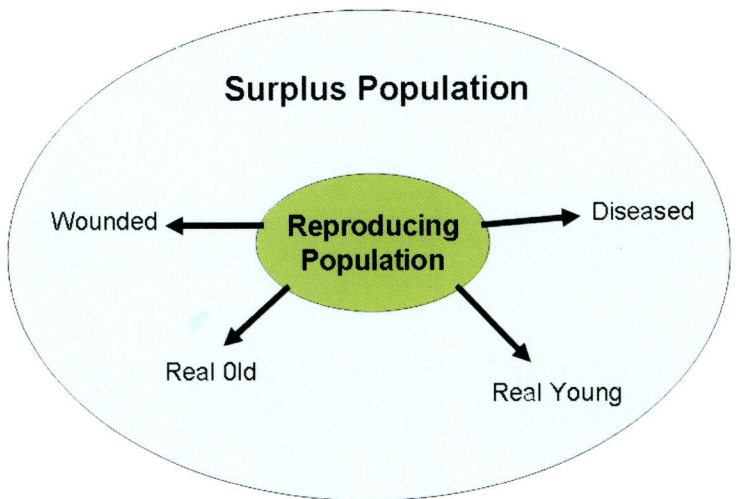

Figure 38. Illustration of how animal populations produce a surplus of members that can be harvested in a controlled and continuous manner (by Clark E. Adams).

In general, a wild population has a reproductive group that produces some "surplus" animals, which can be harvested (fig. 38). The surplus animals consist of young and old, diseased, and wounded population members. The usual result of uncontrolled commercial exploitation of an animal population is to remove both the surplus and the reproductive core, or the entire population. This might be possible at a single rattlesnake den in some cases. However, given decades of hunting and sizes of continuing unregulated harvests of the western diamond-backed rattlesnake in Texas, we appear to have evidence that this animal has a reproductive capacity to maintain a sustainable surplus population.

The level of hunting pressure considered sustainable is called the maximum sustained yield (MSY). At this MSY point, the population has the highest rate of replacement through reproduction. Figure 39 is a graph identifying the MSY point in a population's growth cycle. Below this point, the yield is reduced because the population is small. Above this point, the population is growing so large that it destroys itself (e.g., through density-dependent diseases) and/or destroys its habitat and food sources (e.g., by consuming all of the appropriate prey). MSY is a theoretical concept, however. It is impossible to put real numbers of rattlesnakes at any point on the graph—we do not know the population size, or the

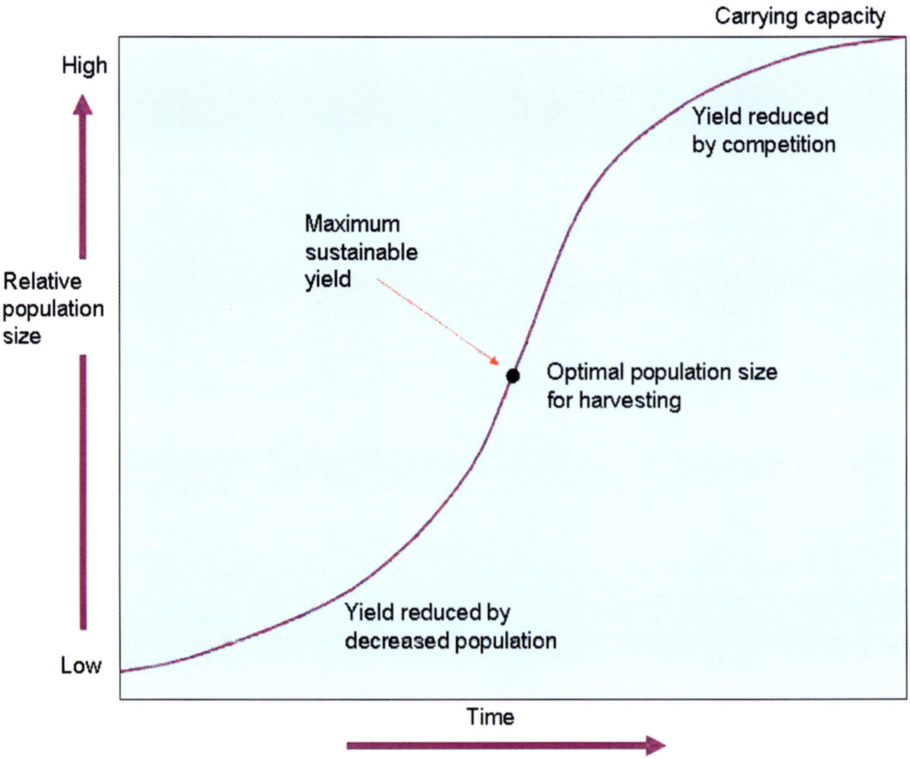

Carrying capacity

High

Yield reduced
by competition

Relative
population
size

Maximum
sustainable
yield

Optimal population size
for harvesting

Yield reduced by
decreased population

Low

Time

Figure 39. Diagram of the concept of maximum sustained yield in animal populations that are being harvested by humans (rendered by Linda Causey from Wright 2005).

carrying capacity of the rattlesnake's habitat, or what point is the actual MSY point. Thus even at the den level of analysis, trying to determine the impact of hunting on rattlesnake populations remains a dilemma.

It is only a slight overstatement to suggest that a researcher would have to spend a lifetime observing western diamond-backed rattlesnakes to derive an answer concerning the impact of hunting on rattlesnake populations in dens. The work would be logistically complex because it

would involve gathering an enormous amount of information over a very long time. Competent field biologists know they cannot conduct studies of this magnitude without adequate research funds and field personnel. Because they face similar questions with other species, wildlife scientists have developed computer models that simulate the habitat conditions required to develop a sustainable harvest of certain animals. These models require a lot of field data (much of it found in published manuscripts on previous field studies); and it requires making some assumptions based on variations in habitat type, sex ratio, age structure, size, life expectancy (i.e., reproductive and mortality rates); and they require density-dependent and independent environmental controls. There is an extensive body of literature on the use of simulation models to predict how variations in selected environmental variables alter the population dynamics of various snake species. However, we found no information that dealt specifically with the impact of hunting on rattlesnake populations.

Chapter 5

Conclusions

Rattlesnake roundups began as a means to rid the land of a perceived threat to people and their livestock. After the 1950s, roundups became larger, more organized events by which rural Texas communities could boost their local economies, attract tourists and their money, and gain notoriety in the national media. During follow-up visits to some of these roundups in 2006, we observed four significant changes that had occurred over the fifteen years since our original 1991 investigation. A few changes were driven by human demographics, while others were a function of market forces. Interestingly, there was no evidence that protests by environmental activists against the harvesting of western diamond-backed rattlesnakes for roundups and commercial trade had had any effect on these changes.

Changes in Rattlesnake Roundups

The first major change was a reduction in the number of roundups in Texas. A roundup requires large investments of time and effort by civic groups and residents. Those roundups that did survive and prosper have dedicated organizers and vast community support. A second important factor is a community's good fortune in being located near major interstate and state highways to accommodate the influx of tourists. Remote communities and their annual roundups were less fortunate given their less accessible locations. Location became ever more important because the state's urban population increased significantly in the 1950s and later, while many rural areas lost population. Curious and paying urban tourists, as well as commercial dealers and vendors, prefer the roundups they can reach on more convenient travel routes. The fact that Sweetwater conducts the largest roundup every year may be linked to its location on Interstate 20, west of Dallas, Fort Worth, and Abilene.

Sweetwater has become synonymous with rattlesnake roundups. As the *Marble Falls Highlander* put it as long ago as 1985, "And in the end, there can only be one." The Sweetwater roundup has long had the highest spectator attendance and is nationally recognized in broadcast and print media. It has a strong community support base among the local residents and the local and national Jaycee organizations, is easily accessed by highway travel, and has hunters bringing thousands of rattlesnakes to the event each year. Rattlesnakes are the main event! In addition the roundup is augmented by a huge entertainment and vendor infrastructure. As noted, Sweetwater realizes an economic windfall of millions of dollars in one weekend. Clearly, the Sweetwater roundup is designed to be a market-driven enterprise. Once you visit the Sweetwater rattlesnake roundup, any other roundup seems anticlimactic. As the number of community-sponsored roundups continues to decline, Sweetwater may become the only rattlesnake roundup left in Texas, with the possible exception of the Freer roundup. Other, smaller roundups continue to wrestle with finding ways to compete with Sweetwater's success. Many communities have given up (see fig. 13).

The second change was the general decline of the rattlesnake trade in Texas and elsewhere. The trade commenced when organizers of roundups discovered that people would pay to attend and would support vendors by buying products made with rattlesnake parts. The market demand for rattlesnakes saw an increase during the 1990s when specialty restaurants in metropolitan areas across the nation began serving exotic meats, including rattlesnake meat. Commercial rattlesnake dealers became the sole source processors and providers of canned rattlesnake meat to these lucrative outlets. That trade has waned somewhat during the past few years. The law of supply and demand has caught up with dealers, who now claim that there are too many rattlesnake skins and too little demand for them. Canned rattlesnake meat was sold regularly at past roundups but not at those visited in 2006. Perhaps the current restaurant demand for rattlesnake meat has declined or restaurateurs have other sources for this exotic menu item.

The commercial trade in rattlesnakes has also changed due to the dealers themselves. Although the number of dealers has always been small, some of the older ones expect to leave their business in the near future, resulting in more consolidation and attrition among this group.

Third, in some communities the methods of attracting public participation to roundups have changed. For example, even though the rattlesnake is still the main attraction at the Sweetwater roundup, the Freer roundup has deemphasized the collection and trade of rattlesnakes. Instcad, thc Freer organizers have focused on providing more alternative forms of family entertainment. The rattlesnake and all the activities surrounding it have become an incidental part of their roundup. Freer could probably have its annual festival without even mentioning rattlesnakes.

Finally, roundups used to be male-dominated events but now have greater participation by women. In past roundups, few women actually hunted snakes; for the most part they participated in beauty contests, cooked the rattlesnake meat, and worked the vendor booths. In 2006 we found that women were just as likely as men to be checking in the snakes collected by hunters, keeping contest records, participating in the education programs, processing the snakes for skins and meat, and performing in the daredevil shows.

Persistence of Opposing Points of View

Over the years, some things did not change. Opposing points of view continue to surface regarding the necessity of and justification for conducting rattlesnake roundups. The arguments for and against roundups can be legitimate, educated or uneducated, limited in application, or downright ludicrous. Consider which of these descriptions best fits the following arguments.

Those who support rattlesnake roundups say:

1. Harvesting rattlesnakes protects livestock and people from fatal rattlesnake bites.

2. Roundups provide outside income for worthy community and civic projects.

3. Roundups educate the public about rattlesnake natural history.

4. Roundups provide venom for medical research regarding the treatment of cancer, heart conditions, tumors, and multiple sclerosis.

5. Roundups provide rattlesnake blood for enzyme research.

6. Roundups provide rattlesnake fat and oil to treat arthritis and rheumatism.

7. Roundups demonstrate the food potential of rattlesnake meat.

8. Roundups have no impact on the 100 million rattlesnakes that inhabit Texas.

9. Roundups generate a market and economic value by the sale of rattlesnake parts—fresh, frozen, fried, and fixed in the form of taxidermy mounts, tanned skins, and other products.

Those who oppose rattlesnake roundups say:

1. Spraying of gasoline into rattlesnake dens affects other species living in the den.

2. Roundups are inhumane.

3. Roundups disrupt the

Point-Counterpoint on Rattlesnake Roundups

On August 6, 2006, newspaperman Donnis Baggett wrote the following article:

Tom Wideman isn't a man to mince words. "A rattlesnake's bite compares to being stabbed with two red-hot ice picks," he writes. Wideman is the former mayor of Sweetwater and former chair of the World's Largest Rattlesnake Roundup, held every March. And now he's a published author, the proud papa of *Texas Rattlesnake Tales*.

Over the past 48 years, Wideman has handled more snakes than a hooker at a political convention. He's as comfortable holding a rattler as a normal man is holding a remote control. His book cover shows him holding up an irate five-foot western diamond-backed rattlesnake with its mouth wide open. Wideman's mouth is wide open, too: He's grinning like a mule eating cactus.

He wasn't always so eager to keep company with snakes. When the Sweetwater Jaycees held their first rattlesnake roundup in 1958, Wideman worked the ticket booth. That was as close to the action as he wanted to be. "I had a strong fear of rattlesnakes and promised myself that I would never hunt or handle the dangerous reptiles," he writes. "In the forty-eight years since, life has taught me many things—chief among them, never say never." Within two years, he was in charge of the roundup's most hazardous duty—weighing and measuring the snakes caught and delivered to the pit.

"Rattlesnakes are generally brought in directly from their dens and haven't been milked, so they're full of venom. Most of them are hot, meaning they're extremely agitated—first at being captured and second at being transported in a container filled with dozens of other ill-tempered rattlesnakes. When containers are opened, rattlesnakes inside strike wildly at anyone and anything within reach." As the years passed, Wideman became one of West Texas' most prolific snake hunters. He's caught countless thousands of the slithery critters—sometimes as many as 300 from a single den—and has shown many a reporter and photographer how it's done.

Along the way, Wideman has appeared on National Geographic Television and has served as a field tester and model for snake proof boots. He estimates his own boots have been struck at least 300 times. But only once has a fang managed to connect with Wideman's flesh. That occurred at his home one day when he was playing with a "pet" rattler named Red Rider. The snake nicked him on the thumb and was promptly evicted into the wild. Wideman recovered fully. "If you're going to adopt rattlesnake hunting as a hobby, the two most important hunting accessories you'll need are a cool head and a deep respect for the critters," Wideman says.

"If you see a rattlesnake or hear one rattle a warning, freeze in your tracks," he warns. "I know that's a difficult assignment when your natural inclination is to run, but you should remain absolutely motionless until you determine the rattlesnake's location. Chances are that if you hear a snake rattle and don't move, the snake will retreat first."

Wideman acknowledges that there's been criticism of the rattlesnake roundup, but he maintains that the annual event has done no long-term damage to the species—or to their dens. He says the dens are repopulated with snakes by the time the next roundup rolls around.

Baggett's piece drew the attention of a professional herpetologist who took strong exception to what had been written about rattlesnakes and roundups. On August 20, 2006, Dr. Lee Fitzgerald wrote the following rebuttal:

Your feature, "Snake Rattle and Roll" (Sunday Aug. 6, 2006), served only the purpose of glorifying Mr. Wideman's exploitation of rattlesnakes and plugging his book, *Texas Rattlesnake Tales*. The "facts" you reported about rattlesnake natural history are at best misleading, but mostly just plain wrong. Both the book and your feature ignore the most important truths about rattlesnake roundups.

As a hunter and outdoorsman, I assure you there is no special glory in rattle-snake hunting. It's not difficult. Wideman was able to collect hundreds of rattle-snakes by spraying gasoline into their dens to drive them out. Rattlesnake roundup organizers turn a blind eye to this unethical hunting practice that not only harms the rattlesnakes themselves, but also the mammals, nonvenomous snakes, lizards, box turtles, and other creatures that co-inhabit the dens.

Too many rattlesnakes are crammed into boxes, barrels, and crates for transport, then dumped into over-crowded snake pits to the point they crush each other to death, defecate on each other, and bite each other (they are not immune to their own bites). The boots Wideman is so proud of are indeed necessary, because roundup organizers stand in the pits kicking and "stirring" the snakes so the ones on the bottom can breathe, and to keep them constantly agitated, rattling, and striking.

The image of jacked-up, threatening, buzzing rattlesnakes has little to do with the natural lives of these secretive, solitary predators. The routine for an adult western diamond-backed rattlesnake is actually very laid back. Rattlesnake venom did not evolve for defense; that's a secondary (but very effective) purpose. Venom is an adaptation for subduing the proportionally large rats, mice, and rabbits that rattlesnakes eat. Believe it or not, an adult rattlesnake may eat only 2 or 3 times a

of such data creates a huge void in our understanding of the magnitude of impacts that commercial exploitation may have on rattlesnake populations.

Although there is a body of popular and professional literature about rattlesnake roundups, most of this information is not common knowledge among the general public. People still need to be made aware of the strategic themes in the published literature. The themes we have addressed, using rattlesnake roundups as a case study, are those of sustainability, sound science, and stewardship. These themes should be considered in the analysis of any human impact on natural habitats and native wildlife.

TPWD and other natural resource agencies are caught between the desires of community organizers to conduct roundups and the concerns of environmental and animal advocacy groups. Agencies are also stymied by budget constraints. As in 1991, the likelihood is that the TPWD director, Wildlife Division head, and commissioners still may not even know how many roundups are still active. Rattlesnake roundups are scarcely a high priority in a tight budget climate.

Recently, the U.S. Department of the Interior, under the State Wildlife Grants Program, appropriated money to all states and territories to assist with initiatives for nongame species. As a requirement for ongoing funding, all states created and published State Wildlife Action Plans. TPWD

published a State Wildlife Action Plan for the period 2005 to 2010. The plan contains:

1. lists of species of special concern
2. the species' conservation priority rating of high, medium, or low
3. any federal designations
4. species abundance rankings
5. associated ecoregion and habitat information
6. problems associated with the species
7. conservation actions
8. monitoring information

Five species of rattlesnake are listed as species of special concern, but the western diamond-backed rattlesnake is not one of the five. In fact, *Crotalus atrox* is not mentioned at all; instead it appears on the "white list" of harvestable species recognized in the nongame permit regulation. One might conclude from this omission that the TPWD considers the statewide population of western diamond-backed rattlesnakes to be sufficiently large and prolific to sustain the continuing removal of tens of thousands of them each year.

Several conditions are needed for a community to realize economic benefits from a rattlesnake roundup. The most critical condition is an abundant and sustainable population of rattlesnakes within close proximity of the roundup location. Next, the rattlesnake has to have economic value. Hunters need to be available to exploit the local rattlesnake population and earn extra income from their efforts. There needs to be a market value for the rattlesnake through commercial dealers, who also expect to make a profit. Taxidermists, who are often the commercial dealers, manufacture trinkets and other marketable products from rattlesnake body parts. They want to make a profit. To best achieve this, rattlesnake roundups would have to continue to be underregulated and undersupervised by the TPWD.

Most important, there must be consumers willing to come to the show and purchase these products. If consumers were eliminated from the market, the whole process would fall apart. The rattlesnake products would gather dust in warehouses; dealers would find alternative employment; and hunters would hunt other wildlife. Even with the elimination of market factors, rattlesnake populations will continue to be adversely affected by other and more deleterious human interventions, such as the loss of habitat due to urban sprawl or other land use practices.

The Importance of Biotic Diversity and Stewardship

What lessons can be learned from the information gathered on Texas rattlesnake roundups? From a histori-cal perspective, rattlesnake hunting today is no different than the market hunting that occurred in North America in the early 1800s. As was the case with game animals then, rattlesnakes today are considered a "commons." The commons refers to a natural resource owned by all, regulated by no one, and with no legal restraints to curb exploitation. In the past, the response of European settlers in North America to wildlife abundance led to the outright extinction of some species (e.g., passenger pigeon) and regional losses of other species (e.g., white-tailed deer, elk, black bear, wild turkey, and many species of waterfowl were drastically reduced across large areas of the East and Midwest in North America). The circumstances resulting in these losses are now referred to as the "tragedy of the commons." The tragedy is measured in terms of how the loss of any species impacts the structure and functioning of an ecosystem. All the animals in an ecosystem have specialized roles to play, called their "niche" in the transfer of energy and nutrients. No species, including humans, can exist by isolating itself from all the other organisms that exist on this planet. It is important to maintain biodiversity in natural ecosystems because all species are interrelated and form an interconnected web of reliance on the existence of one another (see Boeing Analogy sidebar).

Biodiversity can seem a distant and theoretical concept because our understanding of the interconnectedness of the species around us is limited. But its importance is real, and losing biodiversity is hazardous. By applying the accompanying Boeing analogy to natural ecosystems, we can understand how the loss of any species, even the western diamondbacked rattlesnake, reverberates throughout an ecosystem. The repercussions of losing a species can be subtle, but the effects on ecosystem structure and function can also be cataclysmic, depending on the magnitude of the exploitation. The large-scale removal of rattlesnakes from their natural habitat, characteristic of rattlesnake hunting for roundups or for the commercial trade, leaves empty niches that these rattlesnakes once occupied. Repercussions from this loss in terms of ecosystem structure and function might be:

1. loss of a food source for species that feed on rattlesnakes
2. an unnatural break in the predator-prey food web
3. substitution to fill the vacated rattlesnake niche by another

THE BOEING ANALOGY

To understand the value of diversity better, think of an ecosystem as a Boeing 707 jet. The structural components of our 707 ecosystem include sheet metal, rivets, wires, and a host of other different materials (species). Now, imagine you have entered this ecosystem (boarded the plane) and claimed your territory (found a seat). You look through the window and notice that someone is removing rivets (*Rivetus solidus)* from the wing of the plane. You call this to the attention of the flight attendant, who assures you that there is nothing to worry about. There are thousands of rivets in the plane, and the person is just harvesting a few of them to earn a little extra income. You continue your trip without any mishaps.

On your return trip you notice that the same person is removing a few more rivets from the plane. Again you voice your concern to the flight attendant, who again assures you that there are more than enough rivets left. But how many members of the rivet species can be removed before the entire structure collapses?

Perhaps the 707 is a diverse ecosystem with several kinds of fasteners (species) whose function is to hold the plane together. If so, the loss of many or even most rivets may not be disastrous. But what if rivets are the only species performing this function on the 707? What if rivets are a *keystone species,* a part without which the 707 ecosystem cannot exist? Apply this analogy to the ecosystem of your choosing, or even the whole planet, and you will have a good idea of why diversity is so important.

predator that may have a detrimental impact on human endeavors (e.g., coyotes)

4. rapid population growth of prey species in which numbers would otherwise be controlled by the rattlesnake (e.g., rodents)

5. other ecosystem disturbances, understanding of which is still beyond our investigative reach

We expect some readers still to take the view: "Who cares? After all it is just a rattlesnake. The only good rattlesnake is a dead rattlesnake." This attitude represents a failure on our part to understand and respect natural lands and creatures as components of the web of life and failure to recognize the need for "stewardship." Stewardship involves caring for parts of the natural world that are owned by no one but also belong to everyone. Stewardship is the antithesis of the tragedy of the commons. It constitutes recognition of what past generations have left to us and what we may leave to future generations. How sustainability is to be achieved requires our investment of effort.

Regrettably, the prevailing human ethic concerning natural things deals largely with how those things serve human purposes, not how the mere presence of natural things sustains the productivity, order, biodiversity, and life-support systems of our world.

Stewardship cannot be achieved in an intellectual vacuum. It is probably safe to say that few people have more than a vague idea of what services rattlesnakes perform directly or in concert with other species that contribute to a sustainable ecosystem. Some form of instruction on this topic may lead to a different understanding of the importance of the rattlesnake in natural ecosystems, to changed attitudes concerning the animal's value in the wild compared to the marketplace, and to behavioral changes regarding people's use of snakes. Herein lies the foundation for developing the stewardship ethic described by Richard Wright in 2004: "Stewardship becomes a matter of concern that stems from a deep understanding and love of the natural world and the necessary limitations on our use of that world."

Spectator and Hunter Interviews

Students from the Department of Wildlife and Fisheries Sciences at Texas A&M University conducted interviews of spectators and hunters at each roundup. They were assigned to different locations in a roundup facility to maximize access to spectators. Interviewers randomly selected potential respondents from different age, gender, and ethnic groups over the course of six hours each during two days. The questionnaire used was short, resulting in five-minute interviews. Of those spectators asked, 90 percent agreed to participate in an interview. Those who refused did not want to be bothered or, as they said, "did not want to show their ignorance about rattlesnakes."

Spectators

1. What are your reasons for attending this roundup? (Check as identified)

    A. [ ] weekend outing
    B. [ ] view snake contests
    C. [ ] traditional family activity
    D. [ ] learn more about rattlesnakes
    E. [ ] protest rattlesnake roundup
    F. [ ] buy snake products
    G. [ ] other_____

2. How many rattlesnake roundups did you attend last year? _____

3. How far did you travel to come to this roundup? _____ miles

4. From what direction did you travel to come to this roundup?

    A. [ ] N   E. [ ] S
    B. [ ] NE   F. [ ] SW
    C. [ ] E   G. [ ] W
    D. [ ] SE   H. [ ] NW

5. Which of these areas best describe where you live? (Read list and check one only)

    A. [ ] on a farm or ranch
    B. [ ] in open country but not a farm or ranch
    C. [ ] town with < 2,500 people
    D. [ ] town with 2,500 to < 10,000 people
    E. [ ] town with 10,000 to < 50,000 people
    F. [ ] city with 50,000 to < 250,000 people
    G. [ ] city with 250,000 or more people

6. Including this year, how many years have you attended rattlesnake roundups? _____

7. Are you attending this roundup: (Read list and check one only)

    A. [ ] alone
    B. [ ] with another person (unrelated)
    C. [ ] with family—how many? _____
    D. [ ] with organization

8. What effect do you think roundups have on the annual populations of Texas rattlesnakes? (Read list and check one only)

    A. [ ] minimal or no effect
    B. [ ] threatens their future existence
    C. [ ] controls population size
    D. [ ] other: _____

9. What do rattlesnakes eat? (Check as identified)
    A. [ ] insects
    B. [ ] rodents or rabbits
    C. [ ] frogs
    D. [ ] snakes
    E. [ ] birds
    F. [ ] other _____
    G. [ ] do not know

10. Is the presence of rattlesnakes the major reason why you are here? Y [ ] N [ ]

11. Have you attended the educational program at this roundup? Y[ ] N[ ]

Next, spectators were asked to respond to a series of eleven statements by choosing "true," "false," or "do not know." These statements dealt with rattle-snake mythology, natural history, anatomy, and impact on humans and are discussed in the section on Myths about Rattlesnakes.

Hunters

Some students were responsible for interviewing hunters as they brought their rattlesnakes to the check-in point at each roundup. When approached, the hunters rarely refused to participate—less that than one in ten. However, there were times when the interview team was absent when hunters turned in their rattlesnake harvest. Furthermore, it was not possible to estimate the universe of rattlesnake hunters because not every member was present at the roundup weighing-in station where interviews were conducted.

1. Including this year, how many years have you hunted rattlesnakes? _____

2. How many rattlesnake roundups did you attend last year? _____

3. A. How far did you travel to come to this roundup? _____ miles

    B. From what direction did you travel to come to this roundup?

    01. [ ] N   05. [ ] S
    02. [ ] NE   06. [ ] SW
    03. [ ] E   07. [ ] W
    04. [ ] SE   08. [ ] NW

4. Which of these areas best describe where you live? (Read list and check one only)

    A. [ ] on a farm or ranch
    B. [ ] in open country but not a farm or ranch
    C. [ ] town with < 2,500 people
    D. [ ] town with 2,500 to < 10,000 people
    E. [ ] town with 10,000 to < 50,000 people
    F. [ ] city with 50,000 to < 250,000 people
    G. [ ] city with 250,000 or more people

5. Including this year, how many years have you supplied rattlesnakes for roundups?

6. Do you usually hunt rattlesnakes: (Read list and check one only)

    A. [ ] alone
    B. [ ] with a team—how many _____

7. What effect do you think roundups have on the annual populations of Texas rattlesnakes? (Read list and check one only)

    A. [ ] minimal or no effect
    B. [ ] threatens their future existence
    C. [ ] controls population size
    D. [ ] other: _____

8. What do rattlesnakes eat? (Check as identified)

    A. [ ] insects   E. [ ] birds
    B. [ ] rodents/rabbits   F. [ ] other _____
    C. [ ] frogs   G. [ ] do not know
    D. [ ] snakes

9. Would you be in favor of prohibiting the use of gas to capture rattlers?
    Y [ ]   N [ ]

10. For each of the following questions please respond by saying "true" or "false" or "do not know." (Circle response)

    A. T F DN   Rattlesnake age is determined by the number of tail rattles.

    B. T F DN   A rattlesnake is capable of striking at a distance greater than its body length.

    C. T F DN   A rattlesnake is still poisonous if you remove its fangs.

    D. T F DN   A baby rattlesnake's bite is more dangerous than an adult.

    E. T F DN   A rattlesnake sees its prey when it strikes.

    F. T F DN   Livestock losses due to rattlesnake bites are a major problem.

11. How many pounds of rattlesnakes did you or your team bring to this round-up? _____ pounds

12. What is the main reason you hunt rattlesnakes? (Check as identified)

    A. [ ] experience of catching snakes
    B. [ ] additional income
    C. [ ] reduce number of rattlesnakes
    D. [ ] other: _____

13. How many:

    A. times have rattlers bitten you? _____
    B. bites were venomous? _____
    C. venomous bites required a hospital visit? _____

14. Do you provide (sell) snakes to any of the following groups? (Read list and check all that apply)

    A. [ ] dealers or processors
    B. [ ] taxidermists
    C. [ ] pet shops
    D. [ ] folk medicine practitioners
    E. [ ] other roundups
    F. [ ] other:_____

15.   A. In what Texas counties will/have you hunted rattlesnakes this year? (Draw line through counties on reverse side of this page)

    B. In which of these counties do you hunt known rattlesnake dens? (Circle counties on reverse side, if NO dens, skip to 16D)

    C. Do you hunt these dens every year? Y[ ] N[ ]

    D. Within the last two years have you expanded your hunting area into other counties? Y [ ] N [ ]

16. Which of these methods do you use to capture rattlesnakes? (Read list and check all that apply)

    A. [ ] when snakes are warming themselves
    B. [ ] den captures
    C. [ ] road hunt for live snakes
    D. [ ] pick-up road-killed snakes
    E. [ ] other: _____

17. How do you gain access to rattlesnake hunting areas? (Read list and check all that apply)

    A. [ ] landowner request
    B. [ ] land owned by myself/friend/relative
    C. [ ] asking landowner
    D. [ ] advertisements
    E. [ ] other: _____

18. When do you normally hunt rattlesnakes?

    A. [ ] winter only (Do you hunt only for roundups?) Y[ ] N[ ]
    B. [ ] winter through spring
    C. [ ] all year

19. Respondent characteristics: (Used same questions as for spectators)

20. How much do you or your team expect to earn overall from hunting rattlesnakes this year? $_____

antivenin: also known as antivenom, which is a biological commercial product used to treat venomous bites. It is produced by injecting a small amount of a specific venom into a horse, goat, sheep, or rabbit so that the animal's immune system will produce antibodies against the venom. These antibodies are harvested from the animal's blood and processed to treat envenomated victims.

biotic potential: the maximum reproductive potential of a species.

carrying capacity: the maximum number of organisms of any species that can be supported by an ecosystem.

den: any place where rattlesnakes can find refuge from the wind and weather, such as junk piles, haystacks, crawl spaces, and holes between rock layers in a bluff, hill side, or creek bank.

dry bite: a snake bite in which envenomation does not occur.

ecdysis: the process of shedding skin.

ecosystem: a biotic community consisting of a naturally occurring assemblage of organisms interacting with one another and their physical environment.

ectothermia: an organism's ability to regulate its body temperature largely by exchanging heat with its surrounding environment, as is the case with snakes and other reptiles.

envenomation: the injection of venom into an animal by the bite of a venomous animal.

environmental resistance: living (biotic) and nonliving (abiotic) factors that can potentially reduce the population size of a species; living factors include predators, diseases, parasites, and competition, while nonliving factors are unfavorable weather and poor water and food conditions.

fasciotomy: the surgical removal of dead muscle tissue resulting from a venomous snake bite.

game species: any variety of wildlife for which there is an open hunting season for at least part of the year, and normally a big limit, and that is protected at other times.

gassing: the spraying of gasoline into a rattlesnake den to flush snakes out for capture.

hemotoxin: a toxin containing enzymes and other proteins that disrupt blood clotting, destroy red blood cells, and cause organ degeneration and tissue damage; it aids rattlesnakes in the digestion of their prey.

hibernaculum: rock crevices, mammal dens, and other cover where rattlesnakes hibernate.

hibernation: a condition called "winter sleep," whereby some animals are able to slow their metabolism and breathing to a very low level and reduce their body temperature to or close to that of their external environment.

holding pit: a squared or rounded wall-type structure approximately 20 feet across and 4 feet high that is built of Plexiglas, plywood, or concrete to hold several hundred captured western diamond-backed rattlesnakes at a roundup facility.

J-hook: a rod three feet long or longer with a j-shaped curve at one end that is used to control a rattlesnake; also known as a snake hook.

Jacobson's organ: an organ located between the nose and mouth that can detect chemical compounds in the air; a rattlesnake flicks its tongue in the air and touches it to the organ's opening to detect these compounds given off by its prey.

maximum sustainable yield: the maximum harvesting of members of a species that does not lessen the ability of the species to maintain its population size adequately in future generations.

milking: the removal of venom from a rattlesnake by inducing it to bite a collection container.

neurotoxin: a toxin that acts specifically on nerve cells, usually by interacting with cell-membrane proteins to disrupt and destroy the function of live cells.

nongame animals: all wildlife except game mammals, game birds, and aquatic wildlife that can legally be harvested.

ovoviviparous reproduction: the production of eggs that are hatched within the body, so that the young are born alive and without placental attachment.

pancake: describes the coiled, flat position of a rattlesnake; also known as a cow patty.

rattlesnake races: an activity that forces rattlesnakes to proceed down a 20 foot raceway in the shortest time possible.

rattle: a small button of keratin that is produced and remains on the tail each time a western diamond-backed rattlesnake periodically sheds its skin; the sound is produced by a twisting movement of the rattlesnake's tail as it swings from side to side.

rattlesnake roundup: an annually conducted community fund-raising event in which live rattle snakes are collected, commercially traded, and displayed in a fair-like setting.

reproductive rate: how often a species reproduces during a year and how many young are pro-duced.

scalping: the act by rattlesnake dealers offering a better price for live rattlesnakes prior to a roundup event and in comp-etition with the dealer contracted by the roundup organizer to buy all rattlesnakes captured and brought to the roundup facility.

sluff: shed rattlesnake skin.

snake stick: a four- to five-foot-long stick with a clamping device at one end used to grab rattlesnakes safely.

venom: a poison produced by an animal and delivered by means of its fangs, stinger, quills, or barbs.

Adams, C. E., K. J. Lindsey, and S. J. Ash. 2006. *Urban Wildlife Management.* Boca Raton, Fla.: CRC Press, Taylor and Francis Group. 311 pp.

Adams, C. E., J. K. Thomas, K. J. Strnadel, and S. L. Jester. 1994. Texas rattlesnake roundups: Implications of unregulated commercial use of wildlife. *Wildlife Society Bulletin* 22: 324–30.

Adams, C. E., K. J. Strnadel, S. L. Jester, and J. K. Thomas. 1991. *Texas Rattlesnake Roundups.* Contract Rpt. #370-0516. Nongame Species Status Evaluation and Regulation. Texas Parks and Wildlife Department, Austin.

Beavers, R. A. 1976. Food habits of the western diamond-backed rattlesnake, *Crotalus atrox,* in Texas (Viperidae). *Southwestern Naturalist* 20(4): 503–15.

Bush, Sean P. 2004. Snake Envenomations, Rattle. Website: http://www.emedicine.com. Accessed May 10, 2006.

Campbell, J. A., D. R. Fermanowicz Jr., and E. D. Brodie Jr., 1989. Potential impact of rattlesnake roundups on natural populations. *Texas Journal of Science* 41: 301–17.

Charlend, M. B.1989. Size and winter survivorship in neonatal western rattlesnakes, *Crotalus viridis. Canadian Journal of Zoology* 67: 1620–25.

Chase, Alton. 1989. Environment (column), *Austin American-Statesman,* September 14.

Cowles, R. B., and R. L. Phelar. 1958. Olfaction in rattlesnakes. *Copeia* 2: 77–83.

Dixon, James R. 2000. *Amphibians and Reptiles of Texas.* 2nd ed. College Station: Texas A&M University Press.

Etheredge, Clifford. 2004. *Texas Rattlesnake Hunting: One Family's Experience over 100 Years.* Georgetown, Tex.: A Park Imprint. 90 pp.

Fitzgerald, L. A., and C. W. Painter. 2000. Rattlesnake commercialization: Long-term trends, issues, and implications for conservation. *Wildlife Society Bulletin* 28(1): 235–53.

———. 1994. *A Critical Evaluation of Rattlesnake Commercialization: Roundups and Rattlesnake Trade.* Final report to World Wildlife Fund–US/TRAFFIC (USA), Washington, D.C.

Franke, J. 2000. Rattlesnake round-ups: Uncontrolled wildlife exploitation and the rites of spring. *Journal of Applied Animal Welfare Science* 3: 151–60.

French, Rose. 2006. Storms, urban sprawl mean more snakebites. *Bryan–College Station Eagle*, January 15, A10.

Gamow, R. T., and J. E. Harris. 1973. The infrared receptors of snakes. *Scientific American* 228: 94–100.

Gillingham, J. C., and D. L. Clark. 1981. An analysis of prey-searching behavior in the western diamond-backed rattlesnake, *Crotalus atrox*. *Behavioral and Neural Biology* 32: 235–40.

Gloyd, H. K. 1940. *The Rattlesnakes, Genera* Sistrus *and* Crotalus: *A Study in Zoogeography and Evolution*. Chicago: Chicago Academy of Science. 266+ pp.

Graves, B. M. 1989. Defensive behavior of female prairie rattlesnakes (*Crotalus viridis*): Changes after parturition. *Copeia* 3: 791–94.

Griffen, D., and J. W. Donovan. 1986. Significant envenomation from a preserved rattlesnake head (in a patient with a history of immediate hypersensitivity to antivenin). *Annuals of Emergency Medicine* 15: 955–58.

Hardin, G. 1968. The tragedy of the commons. *Science* 162: 1243–48.

http://www.dshs.state.tx.us/idcu/health/zoonosis/animal/bites/information/venom/snake/)

http://www.geocities.com/Baja/Outback/3333/buzz.html.

http://www.geocities.com/RainForest/3096/snakes.html.

http://www.sandiegozoo.org/animalbytes/t-rattlesnake.html.

http://www.desertusa.com/mag98/mar/stories/rattlesin.html.

http://www.texsnakeman.com.

http://whozoo.org/AnlifeSS2001/mindpapr/MP_WesternDiamondback.html.

http://www.fda.gov.

Humane Society. Harmful effects of rattlesnake roundups. Website http://www.hsus.org/wildlife/issues_facing_wildlife. Accessed March 21, 2006.

Kardong, K.V. 1986a. Predatory strike behavior of the rattlesnake, *Crotalus viridis oreganos*. *Journal of Comparative Psychology* 100: 30–14.

Kardong, K.V. 1986b. The predatory strike of the rattlesnake: When things go amiss. *Copeia* 3: 816–20.

Keyler, D. E., and K. Schwitzer. 1987. Envenomation from the fang of a freeze-dried prairie rattlesnake head. *Veterinary and Human Toxicology* 29: 440–41.

Kilmon, Jack, and Hooper Shelton. 1981. *Rattlesnakes in America*. Sweetwater, Tex.: Shelton Press.

King, W. A. Jr., 1964. *Rattling Yours–Snake King.* Published by the author. 223 pp.

Klauber, L. M. 1971. *Rattlesnakes: Their Habits, Life Histories, and Influence on Mankind.* 2 vols. Berkeley: University of California Press.

———. 1972. *Rattlesnakes: Their Habits, Life Histories, and Influence on Mankind.* 2nd ed. 2 vols. Berkeley: University of California Press. 1: xxx + 1–740, 2: xvii + 740–1533.

Lancia, R. A., W. L. Kendall, K. H. Pollock, and J. D. Nichols. 2005. Estimating the number of animals in wildlife populations. Pp. 106–53 in *Techniques for Wildlife Investigations and Management,* ed. C. E. Braun. 6th ed. Bethesda, Md.: Wildlife Society.

Mealer, Bryan. 2000. Jawboning with Snakeburger. Website http://archive.salon.com. Accessed May 31, 2006.

Moon, Brad R., Travis J. LaDuc, Robert Dudley, and Andrew Chang. 2002. A twist to the rattlesnake tail. Pp. 63–76 in *Topics in Functional and Ecological Vertebrate Morphology,* ed. P. Aerts, K.D'Aout, A. Herrel and R. Van Damme. Maastricht, The Netherlands: Shaker Publishing.

Moon, Brad R. J., Johanna Hopp, and Kevin E. Conley. 2002. Mechanical trade-offs explain how performance increases without increasing cost in rattlesnake tail-shaker muscle. *Journal of Experimental Biology* 205: 667–75.

Pough, F. H. 1983. Specializations of the body form and food habits of snakes. *American Zoologist* 23: 443–54.

Price, A. H. 1988. Observations on maternal behavior and neonate aggregation in the western diamond-backed rattlesnake, *Crotalus atrox* (Crotalidae). *Southwestern Naturalist* 33(3): 370–73.

Shelton, H. 1981. A history of the Sweetwater Jaycees Rattlesnake Roundup. Pp. 95–234 in *Rattlesnakes in America,* ed. J. Kilmon and H. Shelton. Sweetwater, Tex.: Shelton Press.

Southwestern Herpetologists Society. 1991. *Herpetology* 21(3). Special edition on Texas rattlesnake roundups.

Speake, D. W., and R. H. Mount. 1973. Some possible ecological effects of "rattlesnake round-ups" in the southeastern coastal plain. *Proceedings of the Annual Conference of the Southeastern Association of Game and Fish Commissioners* 27:267–77.

Strnadel, K. 1993. Hunter participation in rattlesnake roundups. M.S. thesis, Texas A&M University. 84 pp.

Texas Department of State Health Services. 2006. Bites: Venomous Texas Snakes. Website http://www.dshs.state.tx.us. Accessed January 11, 2006.

Thomas, J. K., and C. E. Adams. 1993. The social organization of rattlesnake roundups in rural communities. *Sociological Spectrum* 13: 433–49. http://www.informaworld.com/

Thomas, R. G., and F. H. Pough. 1979. The effect of rattlesnake venom on digestion of prey. *Toxicon* 17: 221–28.

Tinkle, D. W. 1962. Reproductive potential and cycles in female *Crotalus atrox* from northwestern Texas. *Copeia* 2: 306–3.

Warwick, C., C. Steedman, and T. Holford. 1991. Rattlesnake collection drives: Their implications for species and environmental conservation. *Oryx* 25: 39–44.

Weir, Jack. 1992. The Sweetwater rattlesnake round-up: A case study in environmental ethics. *Conservation Biology* 6(1): 116–27.

Werler, J. E., and J. R. Dixon. 2000. *Texas Snakes: Identification, Distribution, and Natural History.* Austin: University of Texas Press.

White, J. 1996. Clinical management of spider bite. *Toxicon* 34:155.

Wideman, Tom. 2006. *Texas Rattlesnake Tales.* Abilene, Tex.: State House Press, McMurry University. 132 pp.

Wright, R. T. 2005. *Environmental Science: Toward a Sustainable Future.* 9th ed. Upper Saddle, N.J.: Prentice Hall.

Wright, A. H., and A. A. Wright. 1957. *Handbook of Snakes of the United States and Canada.* Vol. 2. Ithaca, N.Y.: Comstock Publishing Associates.

Van Riper, W. 1953. How a rattlesnake strikes. *Scientific American* 189(4): 100–102.

Young B. A., C. E. Lee, and K. M. Daley. 2002. Do snakes meter venom? *BioScience* 52(12): 1121–26.

Alldredge, Bud, 20
American Society for Ichthyologists and Herpetologists (ASIH), 29, 55, 84
American Society for the Prevention of Cruelty to Animals, 84
anatomy: head, 4–5; length, 3; rattle, 3, 4, 21; reproduction, 12–14; skin, 1–2, 4
Andrews, 31, 65
Archer City, 31

Baggett, Donnis, 82; "Snake Rattle and Roll," 83, 85
Bibby, Jackie, vi, 65
Big Spring, x, 31, 65
biotic diversity, 29, 87; Boeing analogy, 88, 89; "commons," 87; keystone species, 88; tragedy of the commons, 89
bite: fangs, 5, 6, 21; strike, 6, 21, 49; distance, 6, 21; dog bites, 14, 15; dry bite, 7, 8, 14; envenomation, 5, 7, 22; fasciotomy, 20; human danger, 14, 16, 18, 64, 84; livestock and pet danger, 20, 85; misconception, 7; spider bites, 15; treatment 18, 20, 22, 49. See also first aid; venom
body temperature, 2
Breckenridge, 31
Brownwood, x, 31, 45, 47, 49

Campbell, Jonathan, 54
Center for Disease Control, 84
Chase, Alton, 85
China Spring, 31
Clairemont, 26
Cleburne, 31
cobra, kissing a, 62, 63
Colquitt, Governor O. B., 61
communities, x, 27, 31; Andrews, 31, 65; Archer City, 31; Big Spring, x, 31, 65; Breckenridge, 31; Brownwood, x, 31, 45, 47, 49; China Spring, 31; Clairemont, 26; Cleburne, 31; El Paso, 30; Freer, x, xi, 3, 18, 27, 31, 46, 49, 51, 52, 71, 80, 81; Gainsville, 31; Jacksboro, 31; Lometa, 31; Lubbock, 30; Noack, 30; Oglesby, 31; Okeene (OK), 30; Peducah, 28; San Angelo, 31, 65; San Patricio, 30; Sweetwater, x, xi, 15, 20, 25, 30, 31,32, 35, 36, 38, 40, 41, 43, 46, 47, 49, 51, 58; 64, 65, 68, 79–81; Taylor, x, xi, 18, 21, 30, 31, 38, 43, 44, 51, 81; Weatherford, 31
contests, 44; 41st Annual Peace Officers Rattlesnake Shoot, 26; international races, 16, 45; Miss Snake Charmer, 4, 84; Miss Teen Snake Charmer, 46; sacking contest/championship, 16, 18, 44, 61; stomping contest, 26
copperhead (*Agkistrodon* sp.), xi, 6

daredevil shows, 47, 61, 62, 81
dealers, 3, 13, 33, 67, 68, 72, 80, 87; bids, 34; Bob Popplewell, 34; contracts, 34; Diamondback, 34; George Wills, 34, 59; J. E. Morales, 71; John Shaddix, 34; Maverick, 34; Mike Ivey, 34; Rare Skin, Inc., 34; scalping, 34; Sleepy Leather, 34; Take-a-Memory Home, 34; Tumbleweed Traders, 34. *See also* rattlesnake roundups
dens, xiii, 12, 53, 54, 72, 81, 83; control dens, 74; experimental dens, 74

ecosystem, 8, 10, 88
El Paso, 30
Endangered Species Act, 30
Etheredge, Clifford, 52, 56
eyes, 4, 5

Field of Dreams, 39
first aid, 22; American Red Cross, 20, 22
Fitzgerald, L., and C. W. Painter, 52; Fitzgerald rebuttal to Wideman, 83, 84
folklore, xi; Apostle Paul, 8; Asian folklore, 42; fact or fiction statements, 24; Holiness Movement, 16; myths, 20–24; Saga of Pecos Bill, xi
Freer, x, xi, 3, 18, 27, 31, 46, 49, 51, 52, 71, 80, 81

Gainsville, 31
gasoline fumes, 53–56
gassing/spraying, 37, 53, 54, 56, 70, 81, 83, 85

glossary, 101–103
Greater San Antonio Herpetology Society, 67
Guinness World Record, vi, 65

habitat, 8, 16, 17, 49; biotic potential, 10, 74; carrying capacity, 9, 76; environmental resistance, 10, 74; urbanization, 11. *See also* population, rattlesnake; dens
head, 4, 5
heat-sensing pits, 4, 5
hibernation, 12, 57, 58. *See also* body temperature
Holbrook, Johnnie, 26
holding pit, 40, 41
Humane Society of the United States, The, 29, 84
hunters, 4, 34, 52, 58, 74; access to land, 52; characteristics 53; future limitations, 71; hunting license, 35; number by county, 70; opinions of impacts, 53, 58; personal interviews, x; questionnaire, 96–99; reasons for hunting, 53, 58. *See also* Nongame Species Act; *Texas Rattlesnake Hunting*
hunts, rattlesnake, 53; dens, 53; gassing/spraying, 37, 53, 54, 56, 70, 81, 83, 85; harvest factors, 67; impact on population, 69, 77; J-stick, 53, 54; numbers harvested, 67, 69, 84; patrolling, 56; pounds (biomass) of harvest, 68; storage/transport 57, 58, 81. *See also* gasoline fumes

Jacobson's organ 4, 5
Jacksboro, 31

Kilmon, Jack, 20
Kinsey, Rodney, 38

Lancia, R. A. *See Techniques for
Wildlife Investigations and
Management*
Leopard snake (*Elaphe situal
leopardi*), 8
Liberman, Willie, 61; Snakeville,
61; W. A. Snake King, 61
Lometa, 31
Louv, Richard, xii; *Last Child in the
Woods,* xii; nature deficit disorder,
xii
Lubbock, 30

Marble Falls Highlander, 80

National Geographic Society, 38
National Geographic Television, 81
National Public Radio, 38
Noack, 30
Nongame Species Act, 35; commer-
cial nongame permit, 36, 37, 57,
71

Oglesby, 31
Okeene (OK), 30
opposition, x; ASIH, 29; protest
groups, 36, 38, 49; members,
types of, 37; messages, 37; opponents, 29, 67; opposition views,
81; support views for roundups,
81, 82; spraying opposition,
54–56. *See also* Etheredge; urbanization; biotic diversity; stewardship
O'Reilly, Edward, xii
organizers, 3, 37, 39, 61; bid price,

34; Chamber of Commerce, 39;
contract, 34; Kiwanis, 39; Lions
Club, 39; Sweetwater Jaycees, 38,
39, 70, 80, 81

Peducah, 28
population, rattlesnake, 67, 85; es-
timation of size, 35, 73, 74; field
study of hunting impacts, 72;
irruptive growth, 9; marking procedures, 73; maximum sustained
yield, 75; population crash, 12;
population density, 10; population
growth rate, 10, 89; surplus population, 75
prey, 4, 8; predator-prey balance,
9–12, 88; rodents, 9, 11, 12, 51,
53, 83

Ransberger, Bill, 64, 65
rattle: buttons, 3, 21; function of, 4
rattlesnakes: avoidance, 17; captivity effects, 58; chicken-fried
rattlesnake, 43; "cow patty," 58,
59; ecological service, 8; Eastern diamond-backed (*Crotalus
adamanteus*), 6, 23, 41, 55; "Himalayan rattlesnake" spoof, 59;
Mojave (*Crotalus scutulatus*), 23;
Northern pacific (*Crotalus viridis
oreganos*), 6; range, 27, 28, 71;
Timber (*Crotalus horridus*), 30;
Western diamond-backed (*Crotalus atrox*), ix, xi, xii, 1, 3, 4,
11–14, 17, 18, 23, 27, 35, 41, 44,
46, 50, 56, 57, 60, 63, 68, 70–72,
75, 76, 83, 85, 86
Rattlesnakes in America (Kilmon and
Shelton), 20

reproduction, 12; combat dance, 13; musk glands, 13, 21; ovoviviparous egg development, 13; gestation period, 14; reproductive rate, 74

restaurants, demand by, 29, 43, 80; Café Diablo, 43; Cowboy Club, 43; Rustler's Roost, 43; Seasons Restaurant, 43

roundups, rattlesnake: attendance, 32; changes over time,79–81; daredevil shows, 47, 61, 62, 81; definition, ix; discontinuation, reason for, 29; education programs, 47, 49, 81; entertainment, 26, 46, 62; facilities, 40–47, 58, holding/viewing pit, 2, 40; impacts, 9, 11; landowners, 52; locations, 27; media coverage, 38; music headliners, 46; other states, 25, 29; prices per pound, 33, 71; processing/butchering 41, 42, 81; programs, 39, 47; safety programs, 49; shoots, 25, 26; ticket costs, 32, 61. See also communities; contests; dealers; hunts; hunters; holding pit; organizers; protest groups; vendors

San Angelo, 31, 65
San Patricio, 30
Shelton, Hooper, 20
skin, 2; color patterns, 1, 2; skin shedding, 2; shedding cycle, 4; skin products and prices, 60, 80
snake handlers, 4, 15, 18, 19, 21, 41, 45, 49, 56; clubs, 47, 51, 52, 58; enomaires, 61; Fangs and Rattlers,

61; Heart of Texas Snake Handlers, 61; performances, 61; Rattlers Plus Snake Handlers, 61; reasons for activity, 64; road shows, 50; Sandhills Rattlesnake Club, 61; Snakes Unlimited, 61; South Texas Snake Handlers, 61; Texas State Diamondback Hunters, 61. See also Liberman, Ransberger

snakes, fear of (ophidiophobia), xii
spectators, 49, 50; characteristics, 51; knowledge, 51; misconceptions, 51; opinions of impacts, 52; personal interviews, x, 50; questionnaire, 91–95
stewardship, 86, 87
sustainability, 86, 89; sustainable surplus population, 70, 86. See also population, rattlesnake
Sweetwater, x, xi, 15, 20, 25, 30, 31, 32, 35, 36, 38, 40, 41, 43, 46, 47, 49, 51, 58; 64, 65, 68, 79–81
Sweetwater Reporter, 45

Taylor, x, xi, 18, 21, 30, 31, 38, 43, 44, 51, 81
Techniques for Wildlife Investigations and Management (Lancia, Kendall, Pollock, Nichols), 74
Texas Administrative Code, Chapter 65 on Wildlife, 36, 85
Texas Department of State Health Services, 14
Texas Parks and Wildlife Department, ix, xi, 29, 3, 85–87; Texas Parks and Wildlife Commission, 36
Texas Rattlesnake Hunting (Etheredge), 52

urbanization, 30; urban population, 79; urban sprawl, 87
U.S. Department of the Interior, 86; State Wildlife Grants Program, 86; State Wildlife Action Plans, 86; "white list" of harvestable species, 86
U.S. Fish and Wildlife Service, 30

vendors, 43, 44, 59; products and prices 14, 59, 60. *See also* dealers
venom, 5–7, 83; adult versus young snake, 21; antibodies, 23; anti-

venin, 15, 23, 52, 81; CroFab, 23; effects, 15, 22; extraction equipment, 41; functions, 7; gland, 5; hemotoxicity, 6; milking, 23, 41; neurotoxicity, 6; prices, 41; venomous species, 5, 23. *See also* bite

Wagner, Matt, 36
Weatherford, 31
Wideman, Tom, 82; *Texas Rattlesnake Tales,* 83
Wildlife Society, The, 74
Wright, Richard, 89